Wavelets and Multiwavelets

Studies in Advanced Mathematics

Titles Included in the Series

John P. D'Angelo, Several Complex Variables and the Geometry of Real Hypersurfaces

Steven R. Bell, The Cauchy Transform, Potential Theory, and Conformal Mapping

John J. Benedetto, Harmonic Analysis and Applications

John J. Benedetto and Michael W. Frazier, Wavelets: Mathematics and Applications

Albert Boggess, CR Manifolds and the Tangential Cauchy–Riemann Complex

Goong Chen and Jianxin Zhou, Vibration and Damping in Distributed Systems
    Vol. 1: Analysis, Estimation, Attenuation, and Design
    Vol. 2: WKB and Wave Methods, Visualization, and Experimentation

Carl C. Cowen and Barbara D. MacCluer, Composition Operators on Spaces of Analytic Functions

Jewgeni H. Dshalalow, Real Analysis: An Introduction to the Theory of Real Functions and Integration

Dean G. Duffy, Advanced Engineering Mathematics

Dean G. Duffy, Green's Functions with Applications

Lawrence C. Evans and Ronald F. Gariepy, Measure Theory and Fine Properties of Functions

Gerald B. Folland, A Course in Abstract Harmonic Analysis

José García-Cuerva, Eugenio Hernández, Fernando Soria, and José-Luis Torrea,
    Fourier Analysis and Partial Differential Equations

Peter B. Gilkey, Invariance Theory, the Heat Equation, and the Atiyah-Singer Index Theorem,
    2nd Edition

Peter B. Gilkey, John V. Leahy, and Jeonghueong Park, Spectral Geometry, Riemannian Submersions,
    and the Gromov-Lawson Conjecture

Alfred Gray, Modern Differential Geometry of Curves and Surfaces with Mathematica, 2nd Edition

Eugenio Hernández and Guido Weiss, A First Course on Wavelets

Kenneth B. Howell, Principles of Fourier Analysis

Fritz Keinert, Wavelets and Multiwavelets

Steven G. Krantz, The Elements of Advanced Mathematics, Second Edition

Steven G. Krantz, Partial Differential Equations and Complex Analysis

Steven G. Krantz, Real Analysis and Foundations

Steven G. Krantz, Handbook of Typography for the Mathematical Sciences

Kenneth L. Kuttler, Modern Analysis

Michael Pedersen, Functional Analysis in Applied Mathematics and Engineering

Clark Robinson, Dynamical Systems: Stability, Symbolic Dynamics, and Chaos, 2nd Edition

John Ryan, Clifford Algebras in Analysis and Related Topics

Xavier Saint Raymond, Elementary Introduction to the Theory of Pseudodifferential Operators

John Scherk, Algebra: A Computational Introduction

Pavel Šolín, Karel Segeth, and Ivo Doležel, High-Order Finite Element Method

Robert Strichartz, A Guide to Distribution Theory and Fourier Transforms

André Unterberger and Harald Upmeier, Pseudodifferential Analysis on Symmetric Cones

James S. Walker, Fast Fourier Transforms, 2nd Edition

James S. Walker, A Primer on Wavelets and Their Scientific Applications

Gilbert G. Walter and Xiaoping Shen, Wavelets and Other Orthogonal Systems, Second Edition

Nik Weaver, Mathematical Quantization

Kehe Zhu, An Introduction to Operator Algebras

Wavelets and Multiwavelets

FRITZ KEINERT

CHAPMAN & HALL/CRC

A CRC Press Company
Boca Raton   London   New York   Washington, D.C.

Library of Congress Cataloging-in-Publication Data

Keinert, Fritz.
    Wavelets and multiwavelets / Fritz Keinert.
        p. cm. — (Studies in advanced mathematics)
    Includes bibliographical references and index.
    ISBN 1-58488-304-9 (alk. paper)
    1. Wavelets (Mathematics) I. Title. II. Series.

QA403.3.K45 2003
515′.2433—dc22   2003055780

This book contains information obtained from authentic and highly regarded sources. Reprinted material is quoted with permission, and sources are indicated. A wide variety of references are listed. Reasonable efforts have been made to publish reliable data and information, but the author and the publisher cannot assume responsibility for the validity of all materials or for the consequences of their use.

Neither this book nor any part may be reproduced or transmitted in any form or by any means, electronic or mechanical, including photocopying, microfilming, and recording, or by any information storage or retrieval system, without prior permission in writing from the publisher.

The consent of CRC Press LLC does not extend to copying for general distribution, for promotion, for creating new works, or for resale. Specific permission must be obtained in writing from CRC Press LLC for such copying.

Direct all inquiries to CRC Press LLC, 2000 N.W. Corporate Blvd., Boca Raton, Florida 33431.

Trademark Notice: Product or corporate names may be trademarks or registered trademarks, and are used only for identification and explanation, without intent to infringe.

Visit the CRC Press Web site at www.crcpress.com

© 2004 by Chapman & Hall/CRC

No claim to original U.S. Government works
International Standard Book Number 1-58488-304-9
Library of Congress Card Number 2003055780
Printed in the United States of America 1 2 3 4 5 6 7 8 9 0
Printed on acid-free paper

Preface

Wavelets have been around since the late 1980s, and have found many applications in signal processing, numerical analysis, operator theory, and other fields. Classical wavelet theory is based on a refinement equation of the form

$$\phi(x) = \sqrt{2} \sum_k h_k \phi(2x - k)$$

which defines the scaling function ϕ. The scaling function leads to multiresolution approximations (MRAs), wavelets, and fast decomposition and reconstruction algorithms. Generalizations include wavelet packets, multivariate wavelets, ridgelets, curvelets, vaguelettes, slantlets, second generation wavelets, frames, and other constructions.

One such generalization are multiwavelets, which have been around since the early 1990s. We replace the scaling function ϕ by a function vector

$$\boldsymbol{\phi}(x) = \begin{pmatrix} \phi_1(x) \\ \vdots \\ \phi_r(x) \end{pmatrix}$$

called a *multiscaling function*, and the refinement equation by

$$\boldsymbol{\phi}(x) = \sqrt{m} \sum_k H_k \boldsymbol{\phi}(mx - k).$$

The recursion coefficients H_k are now $r \times r$ matrices.

Multiwavelets lead to MRAs and fast algorithms just like scalar wavelets, but they have some advantages: they can have short support coupled with high smoothness and high approximation order, and they can be both symmetric and orthogonal. They also have some disadvantages: the discrete multiwavelet transform requires preprocessing and postprocessing steps. Also, the theory becomes more complicated.

Many of the existing wavelet books have a short section discussing multiwavelets, but there has never been a full exposition of multiwavelet theory in book form.

This book is divided into two main parts. The first part deals with scalar wavelet theory, and can be read by itself. The second part deals with multiwavelet theory and can also be read by itself, assuming the reader is already familiar with scalar wavelets. Most sections of the two parts run in parallel, so it is easy to refer back and forth between the two and check how a scalar result generalizes to the multiwavelet case. In some cases, the generalization is straightforward. In other cases, the multiwavelet results are more complex, in unexpected ways.

I have chosen to use a dilation factor of 2 in the scalar case, the more general $m \geq 2$ in the multiwavelet case. This way, the first part of the book remains simpler for beginners, and the second part of the book contains more general results. The change from 2 to m has virtually no effect on the difficulty of the material; it just requires a slightly more complicated notation.

I have tried to maintain a balance between mathematical rigor and readability to make the book useful to as wide an audience as possible. The most technical material has been concentrated in two separate chapters that can be skipped without affecting the readability of the other chapters. Most concepts are illustrated with examples.

The book as a whole should be accessible to both mathematicians and engineers, and could be used as the basis for an introductory course or a seminar. Some Matlab routines for experimenting with multiwavelets are available from the author (see appendix C).

Both parts of the book contain the following main topics:

- **Basic theory.** Scaling functions, MRAs, wavelets, moments, approximation order, wavelet decomposition and reconstruction.

- **Practical implementation issues.** Fast algorithms for decomposition and reconstruction, preprocessing, modifications at the boundary, computing point values and integrals.

- **Creating wavelets**

- **Applications.** Signal processing, signal compression, denoising, fast numerical algorithms.

- **Advanced theory.** Existence in the distribution, L^1, L^2, or pointwise sense, stability, smoothness estimates.

The appendix contains a list of standard wavelets, a section on mathematical background, and a list of web and software resources.

Acknowledgments

First and foremost, I would like to express gratitude to my wife Ria, who encouraged me to take on this project and never ceased to lend her encouragement and support. I would also like to thank my editor, Bob Stern, for recruiting and encouraging me. Among the many others who taught me what I know about wavelets, through personal conversation or through their papers, I would like to thank especially Jerry Kautsky, now retired from Flinders University, Adelaide, South Australia, for his hospitality and support during my sabbatical stay and a later visit. I learned a lot of what I know about multiwavelets during that time.

Contents

I   Scalar Wavelets   1

1   Basic Theory   **3**
  1.1   Refinable Functions   3
  1.2   Orthogonal MRAs and Wavelets 10
  1.3   Wavelet Decomposition 19
  1.4   Biorthogonal MRAs and Wavelets 21
  1.5   Moments 25
  1.6   Approximation Order 28
  1.7   Symmetry 32
  1.8   Point Values and Normalization 35

2   Practical Computation   **39**
  2.1   Discrete Wavelet Transform 40
  2.2   Pre- and Postprocessing 43
  2.3   Handling Boundaries 45
        2.3.1   Data Extension Approach 46
        2.3.2   Matrix Completion Approach 48
        2.3.3   Boundary Function Approach 49
        2.3.4   Further Comments 51
  2.4   Putting It All Together 52
  2.5   Modulation Formulation 54
  2.6   Polyphase Formulation 56
  2.7   Lifting 58
  2.8   Calculating Integrals 62
        2.8.1   Integrals with Other Refinable Functions 62
        2.8.2   Integrals with Polynomials 65
        2.8.3   Integrals with General Functions 65

3   Creating Wavelets   **69**
  3.1   Completion Problem 69
        3.1.1   Finding Wavelet Functions 69
        3.1.2   Finding Dual Scaling Functions 70
  3.2   Projection Factors 71
  3.3   Techniques for Modifying Wavelets 72
  3.4   Techniques for Building Wavelets 73

3.5   Bezout Equation 74
3.6   Daubechies Wavelets 75
   3.6.1   Bezout Approach 75
   3.6.2   Projection Factor Approach 78
3.7   Coiflets 79
   3.7.1   Bezout Approach 79
   3.7.2   Projection Factor Approach 81
   3.7.3   Generalized Coiflets 82
3.8   Cohen Wavelets 85
3.9   Other Constructions 86

4 Applications 89
4.1   Signal Processing 89
   4.1.1   Detection of Frequencies and Discontinuities . 90
   4.1.2   Signal Compression 90
   4.1.3   Denoising 91
4.2   Numerical Analysis 93
   4.2.1   Fast Matrix–Vector Multiplication 93
   4.2.2   Fast Operator Evaluation 95
   4.2.3   Differential and Integral Equations 96

5 Existence and Regularity 97
5.1   Distribution Theory 98
5.2 L^1-Theory 100
5.3 L^2-Theory 102
   5.3.1   Transition Operator 102
   5.3.2   Sobolev Space Estimates 105
   5.3.3   Cascade Algorithm 109
5.4   Pointwise Theory 110
5.5   Smoothness and Approximation Order 115
5.6   Stability 116

II Multiwavelets 121

6 Basic Theory 123
6.1   Refinable Function Vectors 124
6.2   MRAs and Multiwavelets 132
   6.2.1   Orthogonal MRAs and Multiwavelets 132
   6.2.2   Biorthogonal MRAs and Multiwavelets 137
6.3   Moments 140
6.4   Approximation Order 142
6.5   Point Values and Normalization 147

Table of Contents xi

7 Practical Computation — 153
- 7.1 Discrete Multiwavelet Transform 154
- 7.2 Pre- and Postprocessing 157
  - 7.2.1 Interpolating Prefilters 158
  - 7.2.2 Quadrature-Based Prefilters 159
  - 7.2.3 Hardin–Roach Prefilters 159
  - 7.2.4 Other Prefilters 161
- 7.3 Balanced Multiwavelets 161
- 7.4 Handling Boundaries 163
  - 7.4.1 Data Extension Approach 163
  - 7.4.2 Matrix Completion Approach 164
  - 7.4.3 Boundary Function Approach 164
- 7.5 Putting It All Together 165
- 7.6 Modulation Formulation 167
- 7.7 Polyphase Formulation 169
- 7.8 Calculating Integrals 171
  - 7.8.1 Integrals with Other Refinable Functions 172
  - 7.8.2 Integrals with Polynomials 174
  - 7.8.3 Integrals with General Functions 175
- 7.9 Applications 175
  - 7.9.1 Signal Processing 175
  - 7.9.2 Numerical Analysis 175

8 Two-Scale Similarity Transforms — 177
- 8.1 Regular TSTs 177
- 8.2 Singular TSTs 179
- 8.3 Multiwavelet TSTs 183
- 8.4 TSTs and Approximation Order 188
- 8.5 Symmetry 191

9 Factorizations of Polyphase Matrices — 197
- 9.1 Projection Factors 197
  - 9.1.1 Orthogonal Case 197
  - 9.1.2 Biorthogonal Case 200
- 9.2 Lifting Steps 203
- 9.3 Raising Approximation Order by Lifting 206

10 Creating Multiwavelets — 209
- 10.1 Orthogonal Completion 209
  - 10.1.1 Using Projection Factors 209
  - 10.1.2 Householder-Type Approach 211
- 10.2 Biorthogonal Completion 212
- 10.3 Other Approaches 214
- 10.4 Techniques for Modifying Multiwavelets 215
- 10.5 Techniques for Building Multiwavelets 216

11 Existence and Regularity   219
    11.1 Distribution Theory 220
    11.2 L^1-Theory 223
    11.3 L^2-Theory 225
        11.3.1 Transition Operator 225
        11.3.2 Sobolev Space Estimates 227
        11.3.3 Cascade Algorithm 231
    11.4 Pointwise Theory 231
    11.5 Smoothness and Approximation Order 235
    11.6 Stability 235

A Standard Wavelets   239
    A.1 Scalar Orthogonal Wavelets 239
    A.2 Scalar Biorthogonal Wavelets 240
    A.3 Orthogonal Multiwavelets 241
    A.4 Biorthogonal Multiwavelets 244

B Mathematical Background   247
    B.1 Notational Conventions 247
    B.2 Derivatives 247
    B.3 Functions and Sequences 247
    B.4 Fourier Transform 249
    B.5 Laurent Polynomials 250
    B.6 Trigonometric Polynomials 250
    B.7 Linear Algebra 252

C Computer Resources   255
    C.1 Wavelet Internet Resources 255
    C.2 Wavelet Software 255
    C.3 Multiwavelet Software 257

References   259

Index   270

Part I

Scalar Wavelets

1

Basic Theory

There are a number of ways to approach wavelet theory. The classic book by Daubechies [50] begins with the continuous wavelet transform and the concept of frames, and does not get to standard wavelet theory until chapter 5. This is more or less the chronological order in which the theory developed.

My personal preference is to begin with the concept of a refinable function and multiresolution approximation (MRA).

1.1 Refinable Functions

DEFINITION 1.1 *A* refinable function *is a function* $\phi : \mathbb{R} \to \mathbb{C}$ *which satisfies a* two-scale refinement equation *or* recursion relation *of the form*

$$\phi(x) = \sqrt{2} \sum_{k=k_0}^{k_1} h_k \, \phi(2x - k). \tag{1.1}$$

The $h_k \in \mathbb{C}$ are called the recursion coefficients.

The refinable function ϕ is called orthogonal *if*

$$\langle \phi(x), \phi(x-k) \rangle = \delta_{0k}, \quad k \in \mathbb{Z}.$$

REMARK 1.2 The number 2 is to wavelet theory what the number 2π is to Fourier theory: it has to be present somewhere, but there is no consensus on where to put it.

If you compare the formulas in this book with those in other books, they may differ by factors of 2 or $\sqrt{2}$. ∎

It is possible to consider refinement equations with an infinite sequence of recursion coefficients. Most of wavelet theory remains valid in this case, as long as the coefficients decay rapidly enough. A typical decay condition is

$$\sum_k |h_k|^{1+\epsilon} < \infty$$

for some $\epsilon > 0$.

Allowing infinite sequences of recursion coefficients requires additional technical conditions in many theorems, and complicates the proof. We will always assume that there are only finitely many nonzero recursion coefficients, which covers most cases of practical interest.

Example 1.1
No book on wavelets is complete without the *Haar function*, which is the characteristic function of the interval $[0, 1]$.

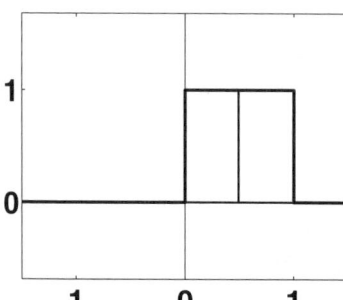
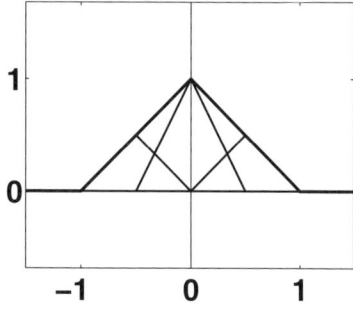

FIGURE 1.1
Left: The Haar function and its refinement equation. Right: The hat function and its refinement equation.

It satisfies
$$\phi(x) = \phi(2x) + \phi(2x - 1) = \sqrt{2}\left(\frac{1}{\sqrt{2}}\phi(2x) + \frac{1}{\sqrt{2}}\phi(2x - 1)\right),$$
so it is refinable with $h_0 = h_1 = 1/\sqrt{2}$ (fig. 1.1).

The Haar function is orthogonal, since for $k \neq 0$ the supports of $\phi(x)$ and $\phi(x - k)$ only overlap at a single point. □

Example 1.2
The *hat function* is given by
$$\phi(x) = \begin{cases} 1 + x & \text{for } -1 \leq x \leq 0 \\ 1 - x & \text{for } 0 < x \leq 1 \\ 0 & \text{otherwise.} \end{cases}$$

Chapter 1: Basic Theory

It satisfies

$$\phi(x) = \frac{1}{2}\phi(2x+1) + \phi(2x) + \frac{1}{2}\phi(2x-1)$$
$$= \sqrt{2}\left(\frac{1}{2\sqrt{2}}\phi(2x+1) + \frac{1}{\sqrt{2}}\phi(2x) + \frac{1}{2\sqrt{2}}\phi(2x-1)\right),$$

so it is refinable with $h_{-1} = h_1 = 1/(2\sqrt{2})$, $h_0 = 1/\sqrt{2}$. It is not orthogonal, since $\langle \phi(x), \phi(x-1)\rangle \neq 0$. □

THEOREM 1.3
A necessary condition for orthogonality is

$$\sum h_k h^*_{k-2\ell} = \delta_{0\ell}. \tag{1.2}$$

The $*$ denotes the complex conjugate transpose, or in the case of a scalar simply the complex conjugate.

PROOF   We calculate

$$\delta_{0\ell} = \langle \phi(x), \phi(x-\ell)\rangle = \langle \sqrt{2}\sum_k h_k\, \phi(2x-k), \sqrt{2}\sum_n h_n\, \phi(2x-2\ell-n)\rangle$$
$$= 2\sum_{k,n} h_k h^*_n \langle \phi(2x-k), \phi(2x-2\ell-n)\rangle$$
$$= \sum_{k,n} h_k h^*_n \delta_{k,2\ell+n} = \sum_k h_k h^*_{k-2\ell}. \qquad\blacksquare$$

The orthogonality condition implies that an orthogonal ϕ has an *even* number of recursion coefficients; otherwise one of the conditions would be $h_{k_0}h^*_{k_1} = 0$.

Many refinable functions are well-defined but cannot be written in closed form. Nevertheless, we can compute values at individual points (or at least approximate them to arbitrary accuracy), compute integrals, determine smoothness properties, and more.

The details will be presented in later chapters. Right now, we just briefly mention two of these techniques, to introduce concepts we need in this chapter.

DEFINITION 1.4   *The* symbol *of a refinable function is the trigonometric polynomial*

$$h(\xi) = \frac{1}{\sqrt{2}} \sum_{k=k_0}^{k_1} h_k e^{-ik\xi}.$$

The Fourier transform of the refinement equation is

$$\hat{\phi}(\xi) = h(\xi/2)\hat{\phi}(\xi/2). \tag{1.3}$$

By substituting this relation into itself repeatedly and taking the limit, we find that formally

$$\hat{\phi}(\xi) = \left[\prod_{k=1}^{\infty} h(2^{-k}\xi)\right]\hat{\phi}(0). \tag{1.4}$$

Assuming that the infinite product converges, this provides a way to compute $\phi(x)$, at least in principle.

We observe that we can choose $\hat{\phi}(0)$ to be an arbitrary number. That is because solutions of refinement equations are only defined up to constant factors: any multiple of a solution is also a solution. The arbitrary factor in equation (1.4) is $\hat{\phi}(0)$. Choosing $\hat{\phi}(0) = 0$ gives $\phi = 0$, which is not an interesting solution; we want $\hat{\phi}(0) \neq 0$.

The infinite product approach is useful for existence and smoothness estimates (see chapter 5), but it is a practical way of finding ϕ only in very simple cases. A better way to get approximate point values of $\phi(x)$ is the *cascade algorithm*, which is fixed point iteration applied to the refinement equation.

We choose a suitable starting function $\phi^{(0)}$, and define

$$\phi^{(n)}(x) = \sqrt{2}\sum_{k} h_k \phi^{(n-1)}(2x-k).$$

This will converge in many cases.

Example 1.3
The scaling function of the Daubechies wavelet D_2 (derived in section 3.6) has recursion coefficients

$$h_0 = \frac{1+\sqrt{3}}{4\sqrt{2}}, \quad h_1 = \frac{3+\sqrt{3}}{4\sqrt{2}}, \quad h_2 = \frac{3-\sqrt{3}}{4\sqrt{2}}, \quad h_3 = \frac{1-\sqrt{3}}{4\sqrt{2}}.$$

It is orthogonal.

Figure 1.2 shows a few iterations of the cascade algorithm for this function.
□

Orthogonality of a refinable function can be checked from the symbol.

THEOREM 1.5
The orthogonality conditions in equation (1.2) are equivalent to

$$|h(\xi)|^2 + |h(\xi+\pi)|^2 = 1. \tag{1.5}$$

These conditions are sufficient to ensure orthogonality if the cascade algorithm for ϕ converges.

Chapter 1: Basic Theory

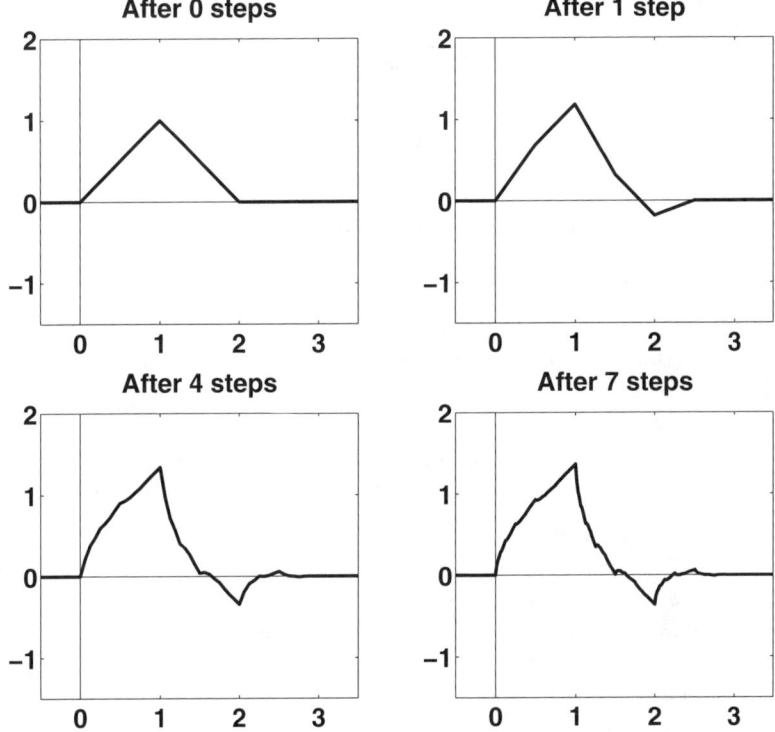

FIGURE 1.2
Cascade algorithm for the scaling function of the Daubechies wavelet D_2. Shown are the starting guess (top left) and the approximate scaling function after one iteration (top right), four iterations (bottom left), and seven iterations (bottom right).

PROOF   We calculate

$$|h(\xi)|^2 + |h(\xi + \pi)|^2 = \frac{1}{2} \sum_{k,n} h_k h_n^* e^{-i(k-n)\xi} \left[1 + (-1)^{k-n}\right]$$

$$= \sum_{k-n \text{ even}} h_k h_n^* e^{-i(k-n)\xi} = \sum_{\ell} \left(\sum_k h_k h_{k-2\ell}^* \right) e^{-2i\ell\xi} = \sum_{\ell} \delta_{0,\ell} e^{-2i\ell\xi} = 1.$$

The sufficiency of these conditions is proved in [50]. The basic idea is simple: we start the cascade algorithm with an orthogonal initial function $\phi^{(0)}$. The orthogonality condition ensures that each $\phi^{(n)}$ will be orthogonal. If the iteration converges, the limit will also be orthogonal. ∎

The *support* of a function ϕ is the closure of the set

$$\{x : \phi(x) \neq 0\}.$$

Compact support means the same as *bounded support*.

LEMMA 1.6
If ϕ is a solution of equation (1.1) with compact support, then

$$\operatorname{supp} \phi = [k_0, k_1].$$

PROOF Assume $\operatorname{supp} \phi = [a, b]$. When we substitute this into the refinement equation, we find that

$$\operatorname{supp} \phi = \left[\frac{a + k_0}{2}, \frac{b + k_1}{2}\right].$$

This implies $a = k_0$, $b = k_1$. ∎

The same argument also shows that if we start the cascade algorithm with a function $\phi^{(0)}$ with support $[a^{(0)}, b^{(0)}]$, the support of $\phi^{(n)}$ will converge to $[k_0, k_1]$ as $n \to \infty$.

For practical applications we need ϕ to have some minimal regularity properties.

DEFINITION 1.7 A refinable function ϕ has **stable shifts** if $\phi \in L^2$ and if there exist constants $0 < A \leq B$ so that for all sequences $\{c_k\} \in \ell^2$,

$$A \sum_k |c_k|^2 \leq \|\sum_k c_k^* \phi(x - k)\|_2^2 \leq B \sum_k |c_k|^2.$$

If ϕ is orthogonal, it is automatically stable with $A = B = 1$. This will be proved in chapter 5 later.

DEFINITION 1.8 A compactly supported refinable function ϕ has **linearly independent shifts** if for all sequences $\{c_k\}$,

$$\sum_k c_k^* \phi(x - k) = 0 \Rightarrow c_k = 0 \quad \text{for all } k.$$

There are no conditions on the sequence $\{c_k\}$ in definition 1.8. Compact support guarantees that for each fixed k the sum only contains finitely many nonzero terms.

Chapter 1: Basic Theory

It can be shown that linear independence implies stability.

REMARK 1.9 You may be wondering at this point why sometimes the coefficients in front of ϕ carry a complex conjugate transpose (as in the definition of stability), and sometimes they do not (as in the refinement equation).

The reason is that I am trying to keep the notation in the scalar wavelet and multiwavelet parts of the book the same as much as possible. At the moment, all coefficients are scalars. In some cases they later turn into row vectors, and in other cases they turn into matrices. It really makes no practical difference at this point whether you add * or not. ∎

THEOREM 1.10
Assume that ϕ is a compactly supported L^2-solution of the refinement equations with nonzero integral and linearly independent shifts. This implies the following conditions:

(i) $h(0) = 1$.

(ii) $\sum_k \phi(x - k) = c, \qquad c \neq 0$ *constant*.

(iii) $h(\pi) = 0$.

PROOF If $\phi \in L^2$ with compact support, it is also in L^1; this implies that $\hat{\phi}$ is continuous and goes to zero at infinity (Riemann–Lebesgue lemma).

Since $\hat{\phi}$ and h are both continuous, equation (1.3) must hold at every point. For $\xi = 0$ we get
$$\hat{\phi}(0) = h(0)\hat{\phi}(0).$$
Since $\hat{\phi}(0) \neq 0$ by assumption, we must have $h(0) = 1$.

By periodicity, $h(2\pi k) = 1$ for any $k \in \mathbb{Z}$. Then
$$\hat{\phi}(4\pi k) = h(2\pi k)\hat{\phi}(2\pi k) = \hat{\phi}(2\pi k),$$
or in general
$$\hat{\phi}(2^n \pi k) = \hat{\phi}(2\pi k), \qquad n \geq 1.$$
Since $|\hat{\phi}(\xi)| \to 0$ as $|\xi| \to \infty$,
$$\hat{\phi}(2\pi k) = 0 \qquad \text{for } k \in \mathbb{Z},\ k \neq 0.$$
By the Poisson summation formula,
$$\sum_k \phi(x-k) = \sqrt{2\pi} \sum_k e^{2\pi i k x} \hat{\phi}(2\pi k) = \sqrt{2\pi}\, \hat{\phi}(0) = \int \phi(x)\, dx = c.$$

Applying the refinement equation to this:

$$c = \sum_k \phi(x-k) = \sqrt{2} \sum_{k,\ell} h_\ell \phi(2x - 2k - \ell)$$

$$= \sum_n \left(\sqrt{2} \sum_k h_{n-2k} \right) \phi(2x - n).$$

By linear independence, we find that

$$\sqrt{2} \sum_k h_{2k} = \sqrt{2} \sum_k h_{2k+1} = 1.$$

Then

$$h(\pi) = \frac{1}{\sqrt{2}} \sum_k h_k e^{-ik\pi} = \frac{1}{\sqrt{2}} \left[\sum_k h_{2k} - \sum_k h_{2k+1} \right] = 0. \quad \blacksquare$$

We will always assume from now on that ϕ satisfies the properties listed in theorem 1.10. We will refer to them as the *basic regularity conditions*.

1.2 Orthogonal MRAs and Wavelets

DEFINITION 1.11 *An multiresolution approximation (MRA) of L^2 is a doubly infinite nested sequence of subspaces of L^2*

$$\cdots \subset V_{-1} \subset V_0 \subset V_1 \subset V_2 \subset \cdots$$

with properties

(i) $\bigcup_n V_n$ *is dense in* L^2.

(ii) $\bigcap_n V_n = \{0\}$.

(iii) $f(x) \in V_n \iff f(2x) \in V_{n+1}$ *for all* $n \in \mathbb{Z}$.

(iv) $f(x) \in V_n \iff f(x - 2^{-n}k) \in V_n$ *for all* $n, k \in \mathbb{Z}$.

(v) *There exists a function* $\phi \in L^2$ *so that* $\{\phi(x-k) : k \in \mathbb{Z}\}$ *forms a stable basis of* V_0.

The basis function ϕ is called the *scaling function*. The MRA is called *orthogonal* if ϕ is orthogonal.

Chapter 1: Basic Theory

Condition (v) means that any $f \in V_0$ can be written uniquely as

$$f(x) = \sum_{k \in \mathbb{Z}} f_k^* \phi(x-k)$$

with convergence in the L^2-sense; and there exist constants $0 < A \le B$, independent of f, so that

$$A \sum_k |f_k|^2 \le \|f\|_2^2 \le B \sum_k |f_k|^2.$$

This implies that ϕ has stable shifts.

Condition (iii) expresses the main property of an MRA: each V_n consists of the functions in V_0 compressed by a factor of 2^n. Thus, a stable basis of V_n is given by $\{\phi_{kn} : n \in \mathbb{Z}\}$, where

$$\phi_{nk}(x) = 2^{n/2} \phi(2^n x - k). \tag{1.6}$$

The factor $2^{n/2}$ preserves the L^2-norm.

Since $V_0 \subset V_1$, ϕ can be written in terms of the basis of V_1 as

$$\phi(x) = \sum_k h_k \phi_{1k}(x) = \sqrt{2} \sum_k h_k \phi(2x - k)$$

for some coefficients h_k. In other words, ϕ is refinable (with possibly an infinite sequence of coefficients). We will assume that the refinement equation is in fact a finite sum.

Let us assume for now that the MRA is orthogonal. The following two lemmas show that in this case ϕ is essentially unique.

LEMMA 1.12
If $\{a_k\}$ is a finite sequence with

$$\sum_k a_k a_{k+\ell}^* = \delta_{0\ell},$$

then

$$a_k = \alpha \delta_{kn}$$

for some α with $|\alpha| = 1$, and some $n \in \mathbb{N}$.

PROOF Let a_{k_0} and a_{k_1} be the first and last nonzero coefficients. Then

$$\delta_{0, k_1 - k_0} = a_{k_0} a_{k_1}^* \ne 0,$$

so we must have $k_0 = k_1 = n$ for some n, and $|a_n|^2 = 1$. ∎

LEMMA 1.13
If ϕ_1 and ϕ_2 are orthogonal scaling functions with compact support which generate the same space V_0, then
$$\phi_2(x) = \alpha \phi_1(x - n)$$
for some constant α with $|\alpha| = 1$, and some $n \in \mathbb{N}$.

PROOF (See [50, chapter 8].) Since ϕ_1 and ϕ_2 are both basis functions of compact support for the same space V_0, we must have
$$\phi_2(x) = \sum_k a_k \phi_1(x - k)$$
for some finite sequence $\{a_k\}$. Orthogonality implies that
$$\sum_\ell a_k a_{k+\ell}^* = \delta_{0\ell},$$
which by lemma 1.12 means
$$a_k = \alpha \delta_{kn}$$
for some $|\alpha| = 1$, $n \in \mathbb{N}$. ∎

The orthogonal projection of an arbitrary function $f \in L^2$ onto V_n is given by
$$P_n f = \sum_k \langle f, \phi_{nk} \rangle \phi_{nk}.$$
The basis functions ϕ_{nk} are shifted in steps of 2^{-n} as k varies, so $P_n f$ cannot represent any detail on a scale smaller than that. We say that the functions in V_n have *resolution* 2^{-n} or *scale* 2^{-n}. $P_n f$ is called an *approximation to f at resolution* 2^{-n}.

An MRA provides a sequence of approximations $P_n f$ of increasing accuracy to a given function f.

LEMMA 1.14
For any $f \in L^2$, $P_n f \to f$ in L^2 as $n \to \infty$.

PROOF Fix f and choose any $\epsilon > 0$. Since $\bigcup_n V_n$ is dense in L^2, we can find an $n \in \mathbb{N}$ and a function $g \in V_n$ so that
$$\|f - g\| < \epsilon.$$
g is automatically in V_k for all $k \geq n$; $P_k f$ is the closest function in V_k to f, so that
$$\|f - P_k f\| \leq \|f - g\| < \epsilon \quad \text{for all } k \geq n. \qquad \blacksquare$$

Chapter 1: Basic Theory

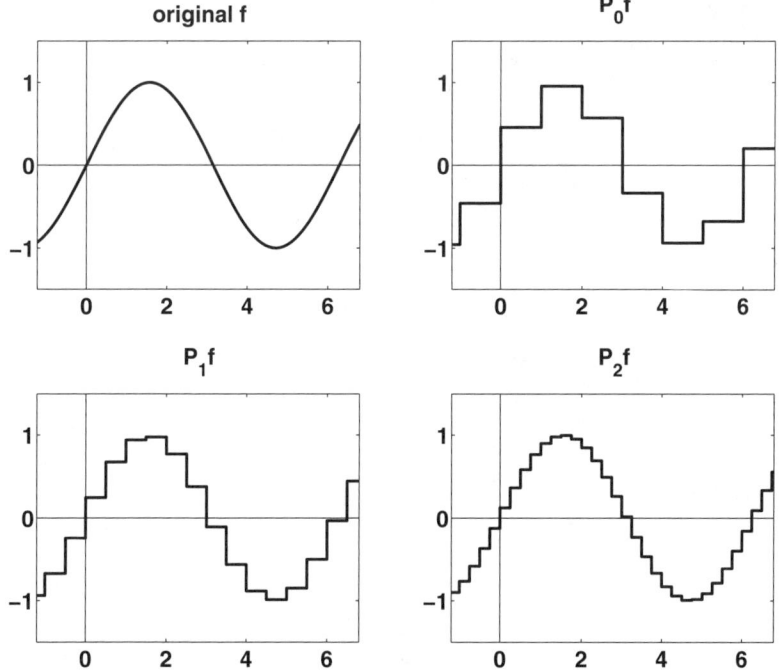

FIGURE 1.3
The function $f(x) = \sin x$ and its approximations $P_0 f$, $P_1 f$, $P_2 f$ at resolutions 1, $1/2$, $1/4$, respectively, based on the Haar function.

Example 1.4
The Haar function produces an orthogonal MRA. Properties (i) and (ii) follow from basic measure theory (step functions are dense in L^2).

V_0 consists of all L^2-functions which are piecewise constant on intervals of the form $[k, k+1)$, $k \in \mathbb{Z}$. V_1 consists of the functions in V_0 compressed by a factor of 2; these are functions piecewise constant on intervals of length $1/2$, of the form $[k/2, (k+1)/2)$. Functions in V_2 are piecewise constant on intervals of length $1/4$, and so on (fig. 1.3). □

The true power of the multiresolution approach arises from considering the differences between approximations at different levels.

The difference between the approximations at resolution 2^{-n} and 2^{-n-1} is called the *fine detail at resolution* 2^{-n}:

$$Q_n f(x) = P_{n+1} f(x) - P_n f(x).$$

Q_n is also an orthogonal projection. Its range W_n is orthogonal to V_n, and

$$V_n \oplus W_n = V_{n+1}.$$

The symbol ⊕ denotes the direct sum of vector spaces (which in this case is an orthogonal direct sum).

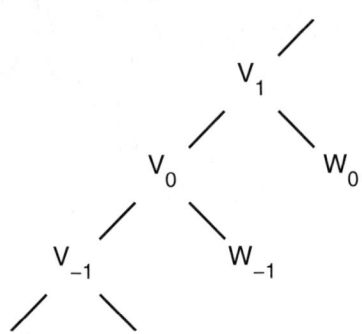

FIGURE 1.4
The spaces V_n and W_n.

The two sequences of spaces $\{V_n\}$ and $\{W_n\}$ and their relationships can be graphically represented as in figure 1.4.

The nesting of the spaces V_n implies that for $k > 0$,

$$P_n P_{n-k} = P_{n-k} P_n = P_{n-k},$$
$$P_n Q_n = 0,$$
$$P_n Q_{n-k} = Q_{n-k}.$$

LEMMA 1.15

$$V_n = \bigoplus_{k=-\infty}^{n-1} W_k.$$

PROOF W_k is a subspace of V_n for all $k < n$, so

$$\bigoplus_{k=-\infty}^{n-1} W_k \subset V_n.$$

Take an arbitrary $f \in V_n$ and decompose it into

$$f = f_n + r$$

with

$$f_n = \sum_{k=-\infty}^{n-1} Q_k f.$$

Chapter 1: Basic Theory

Then

$$P_{n-1}f_n = P_{n-1}Q_{n-1}f + \sum_{k=-\infty}^{n-2} P_{n-1}Q_k f$$

$$= \sum_{k=-\infty}^{n-2} Q_k f = f_{n-1},$$

and

$$\begin{aligned} f = P_n f &= P_{n-1}f + Q_{n-1}f \\ &= P_{n-1}f_n + P_{n-1}r + Q_{n-1}f \\ &= f_{n-1} + P_{n-1}r + Q_{n-1}f \\ &= f_n + P_{n-1}r, \end{aligned}$$

so

$$P_{n-1}r = r.$$

We can prove likewise that $P_k r = r$ for all $k < n$, so

$$r \in \bigcap_{k=-\infty}^{n-1} V_k = \{0\}. \qquad \blacksquare$$

The sequence of spaces $\{W_n\}$ satisfies conditions similar to conditions (i) through (v) of an MRA. The symbol $\perp$ stands for "is orthogonal to."

THEOREM 1.16

For any orthogonal MRA with scaling function ϕ,

(i) $\bigoplus_n W_n$ *is dense in L^2.*

(ii) $W_k \perp W_n$ *if $k \neq n$.*

(iii) $f(x) \in W_n \iff f(2x) \in W_{n+1}$ *for all $n \in \mathbb{Z}$.*

(iv) $f(x) \in W_n \iff f(x - 2^{-n}k) \in W_n$ *for all $n, k \in \mathbb{Z}$.*

(v) *There exists a function $\psi \in L^2$ so that $\{\psi(x-k) : k \in \mathbb{Z}\}$ forms an orthogonal stable basis of W_0, and $\{\psi_{nk} : n, k \in \mathbb{Z}\}$ forms a stable basis of L^2.*

(vi) *Since $\psi \in V_1$, it can be represented as*

$$\psi(x) = \sqrt{2} \sum_k g_k \phi(2x - k)$$

for some coefficients g_k. If h_k are the recursion coefficients of ϕ, then we can choose

$$g_k = (-1)^k h_{N-k}, \tag{1.7}$$

where N is any odd number.

The function ψ is called the wavelet function or mother wavelet. ϕ and ψ together form a wavelet.

PROOF   Property (i) follows from lemma 1.15 and property (i) of an MRA:

$$\bigcup_{k=-\infty}^{n} V_k = V_n = \bigoplus_{k=-\infty}^{n-1} W_k.$$

Property (ii) is easy to check: without loss of generality we can assume $k < n$. Then $W_k \subset V_{k+1} \subset V_n$, and $V_n \perp W_n$.

Properties (iii) and (iv) are inherited from the corresponding properties of an MRA.

Formula (1.7) for ψ will be derived in section 3.1.1, where we will also show that this is essentially the only possible choice.

The number N is usually taken to be $k_0 + k_1$ (which is always odd). In this case, ϕ and ψ have the same support.

Stability of ψ will be shown in chapter 5. ∎

Note that ψ is not a refinable function: it is defined in terms of ϕ, not in terms of itself.

Example 1.5
By equation (1.7), the wavelet function for the Haar scaling function (fig. 1.5) is given by the coefficients

$$\{g_0, g_1\} = \{h_1, -h_0\} = \{\frac{1}{\sqrt{2}}, -\frac{1}{\sqrt{2}}\}.$$

Since $\phi = \chi_{[0,1]}$, we get

$$\psi = \chi_{[0,1/2]} - \chi_{[1/2,1]}.$$

Likewise, the wavelet function for the Daubechies wavelet D_2 (fig. 1.6) is a linear combination of four compressed and shifted versions of ϕ. It is given by the coefficients

$$\{g_0, g_1, g_2, g_3\} = \{h_3, -h_2, h_1, -h_0\}$$
$$= \left\{\frac{1-\sqrt{3}}{4\sqrt{2}}, -\frac{3-\sqrt{3}}{4\sqrt{2}}, \frac{3+\sqrt{3}}{4\sqrt{2}}, -\frac{1+\sqrt{3}}{4\sqrt{2}}\right\}. \quad □$$

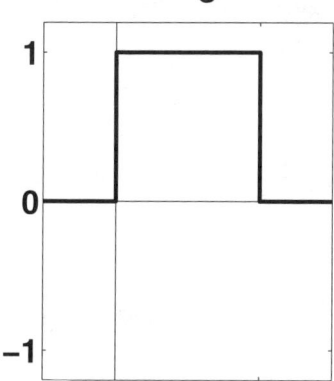
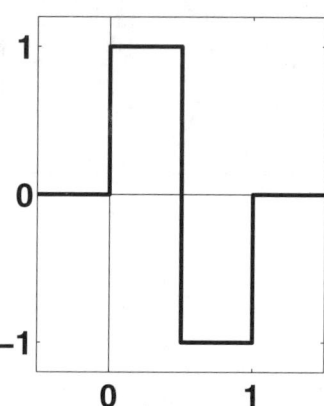

FIGURE 1.5
The Haar scaling function (left) and wavelet (right).

We define the symbol of ψ as

$$g(\xi) = \frac{1}{\sqrt{2}} \sum_k g_k e^{-ik\xi}.$$

The Fourier transform of the refinement equation for ψ is

$$\hat{\psi}(\xi) = g(\xi/2)\hat{\phi}(\xi/2).$$

Orthogonality of ϕ and ψ can be expressed as in theorem 1.3

$$\sum h_k h_{k-2\ell}^* = \sum g_k g_{k-2\ell}^* = \delta_{0\ell},$$
$$\sum h_k g_{k-2\ell}^* = \sum g_k h_{k-2\ell}^* = 0, \tag{1.8}$$

or equivalently

$$|h(\xi)|^2 + |h(\xi+\pi)|^2 = |g(\xi)|^2 + |g(\xi+\pi)|^2 = 1,$$
$$h(\xi)g(\xi)^* + h(\xi+\pi)g(\xi+\pi)^* = g(\xi)h(\xi)^* + g(\xi+\pi)h(\xi+\pi)^* = 0. \tag{1.9}$$

In terms of the wavelet function, the projection Q_n is given by

$$Q_n f = \sum_k \langle f, \psi_{nk} \rangle \psi_{nk}.$$

We now come to the main concept we seek: the *discrete wavelet transform* (DWT).

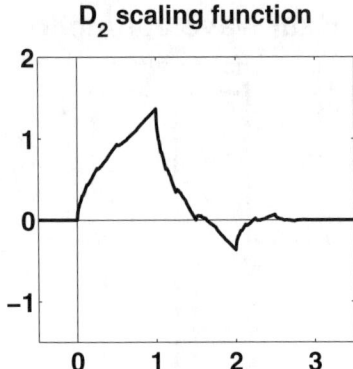 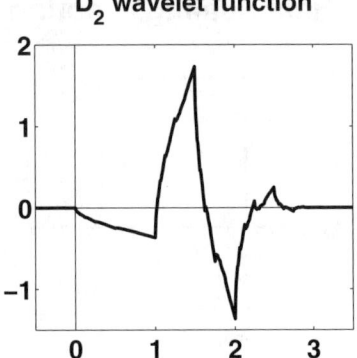

FIGURE 1.6
The scaling function (left) and wavelet function (right) of the Daubechies wavelet D_2.

Given a function $f \in L^2$, we can represent it as

$$f = \sum_{k=-\infty}^{\infty} Q_k f$$

(complete decomposition in terms of detail at all levels). Alternatively, we can start at any level ℓ and use the approximation at resolution $2^{-\ell}$ plus all the detail at finer resolution:

$$f = P_\ell f + \sum_{k=\ell}^{\infty} Q_k f.$$

For practical applications, we need to reduce this to a finite sum. We assume that $f \in V_n$ for some $n > \ell$. Then

$$f = P_n f = P_\ell f + \sum_{k=\ell}^{n-1} Q_k f. \tag{1.10}$$

Equation (1.10) describes the DWT: the original function or signal f gets decomposed into a coarse approximation $P_\ell f$, and fine detail at several resolutions (fig. 1.7). The decomposition as well as the reconstruction can be performed very efficiently on a computer. Implementation details are presented in section 2.1.

Chapter 1: Basic Theory

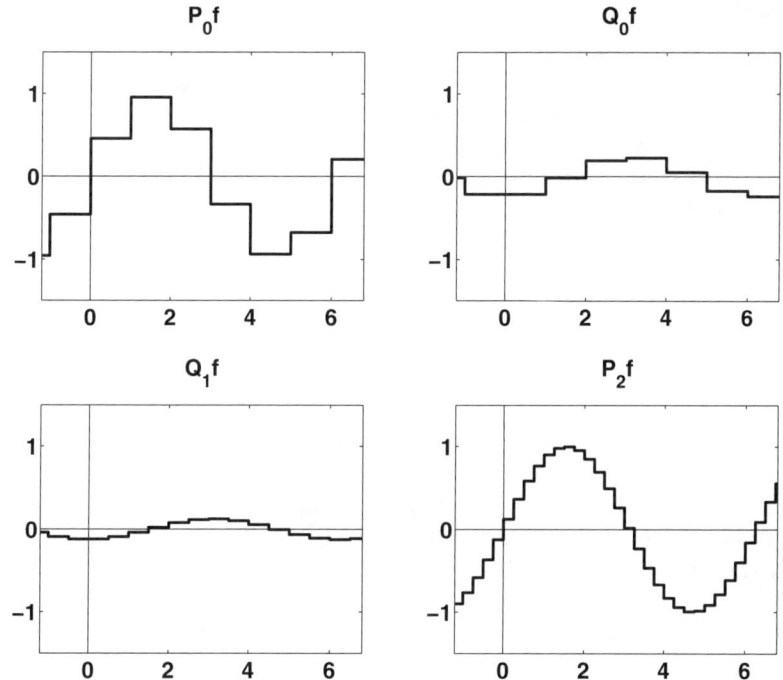

FIGURE 1.7
$P_0 f$ (top left), $Q_0 f$ (top right), $Q_1 f$ (bottom left), and $P_2 f = P_0 f + Q_0 f + Q_1 f$ (bottom right) for $f(x) = \sin x$.

1.3 Wavelet Decomposition

We have already explained the basic idea behind the decomposition of a function in terms of different scales. This section describes a different way of looking at the same idea. It is more heuristic than mathematical in nature.

The Fourier transform is a *frequency* transform: if $f(t)$ is a function of time, $\hat{f}(\xi)$ is interpreted as the frequency content of f at frequency ξ. This works well for detecting the main frequencies in a signal, but does not give an easy way to find out *when* in time the frequencies are present.

For example, given a recording of three notes played on a piano, one could easily determine from the Fourier transform what the three notes were. However, it would be hard to determine in which order the notes were played.

Ideally, we would like to have a *time-frequency transform* F of f where $F(t, \xi)$ represents the frequency content of f at frequency ξ at time t. Such an F cannot actually exist: in order to detect a frequency, you have to observe the signal for a period of time. That is the uncertainty principle. However, it is possible to determine approximate local time-frequency averages.

If **u** is a unit vector, **x** an arbitrary vector in $\mathbb{R}^n$, the inner product $\langle \mathbf{x}, \mathbf{u} \rangle$ is the size of the component of **x** in direction **u**. Likewise if $f, w \in L^2(\mathbb{R})$ with $\|w\|_2 = 1$, we can interpret $\langle f, w \rangle$ as the component of f in direction w, that is, as a measure of how much f "looks like" w. It is simultaneously a measure of how much $\hat{f}$ looks like $\hat{w}$, since by Parseval's formula

$$\langle f, w \rangle = \langle \hat{f}, \hat{w} \rangle.$$

Fix a particular function $w \in L^2$ with $\|w\|_2 = 1$. We assume that

$$\mu = \int t |w(t)|^2 \, dt,$$

$$\sigma = \left\{ \int (t - \mu)^2 |w(t)|^2 \, dt \right\}^{1/2}$$

exist, along with the corresponding quantities $\hat{\mu}$, $\hat{\sigma}$ of $\hat{w}$. These are the mean and standard deviation of the probability distributions $|w|^2$ and $|\hat{w}|^2$. w is approximately localized in the time interval $\mu - \sigma \leq t \leq \mu + \sigma$ and frequency interval $\hat{\mu} - \hat{\sigma} \leq \xi \leq \hat{\mu} + \hat{\sigma}$.

The inner product $\langle f, w \rangle$ can then be interpreted as the frequency content of f in the time interval $\mu - \sigma \leq t \leq \mu + \sigma$ and frequency interval $\hat{\mu} - \hat{\sigma} \leq \xi \leq \hat{\mu} + \hat{\sigma}$, or equivalently as a local average of the (hypothetical) time-frequency distribution F over the corresponding box in time-frequency space.

The uncertainty principle

$$\sigma \cdot \hat{\sigma} \geq 1/2$$

requires this box to have a minimum area of 2, but the shape and location can be controlled by choosing a suitable w. We can have good time resolution (small σ) at the cost of bad frequency resolution (large $\hat{\sigma}$), or vice versa. f and $\hat{f}$ represent the extremes: f has perfect time resolution, no frequency resolution; $\hat{f}$ has perfect frequency resolution, no time resolution. Other methods, including the wavelet decomposition, are somewhere in between.

One alternative approach to an approximate time-frequency transform is the short-term Fourier transform (STFT). We fix a window function w, choose time and frequency steps Δt, $\Delta \xi$, and compute the coefficients

$$f_{nk} = \langle f, w_{nk} \rangle,$$

where

$$w_{nk}(t) = e^{int\Delta\xi} w(t - k\Delta t).$$

Each w_{nk} corresponds to a box in time-frequency space. All boxes have the same size $(2\sigma) \times (2\hat{\sigma})$, and their centers form a rectangular grid with spacings Δt, $\Delta \xi$. The spacings should satisfy

$$\Delta t \leq 2\sigma,$$

$$\Delta \xi \leq 2\hat{\sigma},$$

Chapter 1: Basic Theory 21

to make sure the entire plane gets covered. The STFT provides the same time and frequency resolution everywhere.

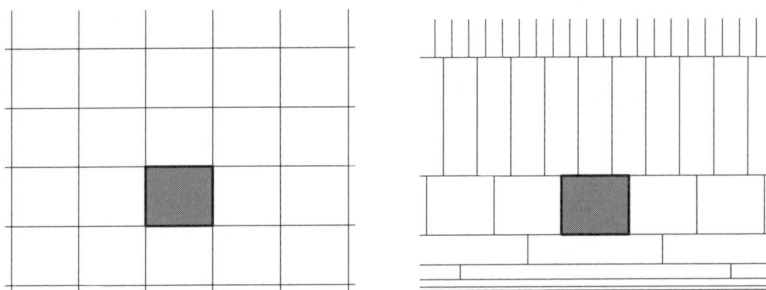

FIGURE 1.8
Tiling of the time-frequency plane for STFT (left) and DWT (right). The time axis runs left to right, the frequency axis bottom to top.

The DWT uses an analogous approach with a mother wavelet ψ, and

$$\psi_{nk}(t) = 2^{n/2}\psi(2^n t - k).$$

Here the nk-box has size $2^{-n}(2\sigma) \times 2^n(2\hat{\sigma})$, with different spacings at different frequencies (fig 1.8).

The DWT provides different time and frequency resolutions at different frequencies. For low frequencies, we get good frequency resolution, but bad time resolution. For high frequencies it is the other way around.

This approach works well in practice. In analyzing a sound recording, the onset of a sound is often accompanied by high-frequency transients which die out rapidly, in a time proportional to 1/frequency. Lower frequencies persist for longer times. That is precisely the scale on which the DWT tries to capture them.

1.4 Biorthogonal MRAs and Wavelets

Refinable functions that define MRAs are relatively easy to find. Orthogonal MRAs are harder to find, but the orthogonality requirement can be replaced by milder biorthogonality conditions.

DEFINITION 1.17 *Two refinable functions ϕ, $\tilde{\phi}$ are called* **biorthogonal**

if
$$\langle \phi(x), \tilde{\phi}(x-k) \rangle = \delta_{0k}.$$
We also call $\tilde{\phi}$ the dual of ϕ.

THEOREM 1.18
A necessary condition for biorthogonality is
$$\sum h_k \tilde{h}_{k-2\ell}^* = \delta_{0\ell}, \qquad (1.11)$$
or equivalently
$$h(\xi)\tilde{h}(\xi)^* + h(\xi+\pi)\tilde{h}(\xi+\pi)^* = 1. \qquad (1.12)$$

These conditions are sufficient to ensure biorthogonality if the cascade algorithm for both ϕ and $\tilde{\phi}$ converges.

The proof is analogous to that of theorem 1.3.

REMARK 1.19 It is possible to orthonormalize an existing scaling function with stable shifts [109], but the resulting new ϕ does not usually have compact support any more. This makes it less desirable for practical applications. ■

Assume now that we have two MRAs $\{V_n\}$ and $\{\tilde{V}_n\}$, generated by biorthogonal scaling functions ϕ and $\tilde{\phi}$. We can complete the construction of wavelets as follows.

The projections P_n and $\tilde{P}_n$ from L^2 into V_n, $\tilde{V}_n$, respectively, are given by
$$P_n f = \sum_k \langle f, \tilde{\phi}_{nk} \rangle \phi_{nk},$$
$$\tilde{P}_n f = \sum_k \langle f, \phi_{nk} \rangle \tilde{\phi}_{nk},$$
where ϕ_{nk}, $\tilde{\phi}_{nk}$ are defined as in equation (1.6). These are now oblique (i.e., nonorthogonal) projections.

The projections Q_n, $\tilde{Q}_n$ are defined as before by
$$Q_n f = P_{n+1} f - P_n f,$$
$$\tilde{Q}_n f = \tilde{P}_{n+1} f - \tilde{P}_n f,$$
and their ranges are the spaces W_n, $\tilde{W}_n$.

The space W_n is orthogonal to $\tilde{V}_n$: if $f \in W_n$, then
$$f = Q_n f = P_{n+1} f - P_n f.$$

Chapter 1: Basic Theory

However, $W_n \subset V_{n+1}$, so $f = P_{n+1}f$ and $P_n f = 0$. This means $\langle f, \tilde{\phi}_{nk}\rangle = 0$ for all k, or $f \perp \tilde{V}_n$.

We still have
$$V_n \oplus W_n = V_{n+1}$$
as a nonorthogonal direct sum.

The question is now whether we can find wavelet functions $\psi, \tilde{\psi}$ which span the spaces W_n, $\tilde{W}_n$. Finding the functions is not hard. The hard part is the stability.

THEOREM 1.20
Assume that $\phi, \tilde{\phi} \in L^2$ are scaling functions generating biorthogonal MRAs, and that the cascade algorithm converges for both of them. Then

(i) $\bigoplus_n W_n$, $\bigoplus_n \tilde{W}_n$ *are dense in* L^2.

(ii) $W_k \perp \tilde{W}_n$ *if* $k \neq n$.

(iii)
$$f(x) \in W_n \iff f(2x) \in W_{n+1} \quad \text{for all } n \in \mathbb{Z},$$
$$f(x) \in \tilde{W}_n \iff f(2x) \in \tilde{W}_{n+1} \quad \text{for all } n \in \mathbb{Z}.$$

(iv)
$$f(x) \in W_n \iff f(x - 2^{-n}k) \in W_n \quad \text{for all } n, k \in \mathbb{Z},$$
$$f(x) \in \tilde{W}_n \iff f(x - 2^{-n}k) \in \tilde{W}_n \quad \text{for all } n, k \in \mathbb{Z}.$$

(v) There exist biorthogonal functions $\psi, \tilde{\psi} \in L^2$ so that $\{\psi(x-k) : k \in \mathbb{Z}\}$ forms a stable basis of W_0, $\{\tilde{\psi}(x-k) : k \in \mathbb{Z}\}$ forms a stable basis of $\tilde{W}_0$, and $\{\psi_{nk} : n, k \in \mathbb{Z}\}$, $\{\tilde{\psi}_{nk} : n, k \in \mathbb{Z}\}$ both form a stable basis of L^2.

(vi) Since $\psi \in V_1$, $\tilde{\psi} \in \tilde{V}_1$, they can be represented as
$$\psi(x) = \sqrt{2}\sum_k g_k \phi(2x-k),$$
$$\tilde{\psi}(x) = \sqrt{2}\sum_k \tilde{g}_k \tilde{\phi}(2x-k)$$

for some coefficients $g_k, \tilde{g}_k$.

If $h_k, \tilde{h}_k$ are the recursion coefficients of $\phi, \tilde{\phi}$, we can choose
$$\begin{aligned}\psi(x) &= \sqrt{2}\sum_k (-1)^k \tilde{h}^*_{N-k} \phi(2x-k), \\ \tilde{\psi}(x) &= \sqrt{2}\sum_k (-1)^k h^*_{N-k} \tilde{\phi}(2x-k),\end{aligned} \quad (1.13)$$

where N is any odd number.

The functions ψ, $\tilde{\psi}$ are again called the wavelet functions or mother wavelets.

The proof is basically the same as that of theorem 6, except for the stability part. That will be covered in chapter 5 (theorem 5.27).

As in equations (1.2) and (1.5), we can express the biorthogonality conditions as

$$\sum h_k \tilde{h}^*_{k-2\ell} = \sum g_k \tilde{g}^*_{k-2\ell} = \delta_{0\ell},$$
$$\sum h_k \tilde{g}^*_{k-2\ell} = \sum g_k \tilde{h}^*_{k-2\ell} = 0,$$
(1.14)

or equivalently

$$h(\xi)\tilde{h}(\xi)^* + h(\xi+\pi)\tilde{h}(\xi+\pi)^* = g(\xi)\tilde{g}(\xi)^* + g(\xi+\pi)\tilde{g}(\xi+\pi)^* = 1,$$
$$h(\xi)\tilde{g}(\xi)^* + h(\xi+\pi)\tilde{g}(\xi+\pi)^* = g(\xi)\tilde{h}(\xi)^* + g(\xi+\pi)\tilde{h}(\xi+\pi)^* = 0.$$
(1.15)

REMARK 1.21 The dual scaling function is not unique. A given scaling function may not have any dual. If one dual exists, however, dual lifting steps (section 2.7) will produce an infinite number of others. ∎

Example 1.6
One of many possible scaling functions dual to the hat function is defined by the recursion coefficients

$$\{\tilde{h}_{-4},\ldots,\tilde{h}_4\} = \frac{\sqrt{2}}{128}\{3,-6,-16,38,90,38,-16,-6,3\}.$$

This pair is called the Cohen(2,4) wavelet (fig. 1.9).

The corresponding wavelet functions have coefficients

$$\{g_{-3},\ldots,g_5\} = \frac{\sqrt{2}}{128}\{-3,-6,16,38,-90,38,16,-6,-3\},$$
$$\{\tilde{g}_0,\tilde{g}_1,\tilde{g}_2\} = \frac{\sqrt{2}}{4}\{1,-2,1\}.$$

We can use equations (1.14) to verify that these coefficients work. We will see in section 3.8 where they come from. ☐

As before, $P_n f$ is interpreted as an approximation to f at resolution 2^{-n}, and $Q_n f$ is the fine detail. If $f \in V_n$, we can do a wavelet decomposition

$$f = P_n f = P_\ell f + \sum_{k=\ell}^{n-1} Q_k f.$$

Chapter 1: Basic Theory

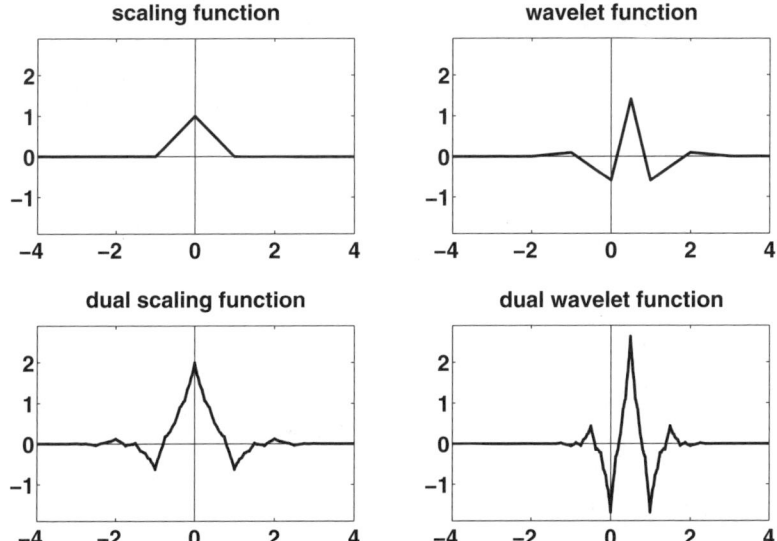

FIGURE 1.9
Scaling functions (left) and wavelet functions (right) for the Cohen(2,4)-wavelet (top) and its dual (bottom).

However, we can also do these things on the dual side. $\tilde{P}_n f$ is also an approximation to f at resolution 2^{-n}, with fine detail $\tilde{Q}_n f$. If $f \in \tilde{V}_n$, we can do a dual wavelet decomposition

$$f = \tilde{P}_n f = \tilde{P}_\ell f + \sum_{k=\ell}^{n-1} \tilde{Q}_k f.$$

This is an example of the general symmetry about biorthogonal wavelets: whenever you have any formula or algorithm, you can put a tilde on everything that did not have one, and vice versa, and you get a dual formula or algorithm.

That does not mean that both algorithms are equally useful for a particular application, but they both work.

1.5 Moments

We always assume that ϕ satisfies the minimal regularity assumptions from theorem 1.10.

DEFINITION 1.22 *The kth discrete moments of ϕ, ψ are defined by*

$$m_k = \frac{1}{\sqrt{2}} \sum_\ell \ell^k h_\ell,$$

$$n_k = \frac{1}{\sqrt{2}} \sum_\ell \ell^k g_\ell.$$

They are related to the symbols by

$$\begin{aligned} m_k &= i^k D^k h(0), \\ n_k &= i^k D^k g(0). \end{aligned} \qquad (1.16)$$

In particular, $m_0 = h(0) = 1$.

Discrete moments are uniquely defined and easy to calculate.

DEFINITION 1.23 *The kth continuous moments of ϕ, ψ are*

$$\begin{aligned} \mu_k &= \int x^k \phi(x)\, dx, \\ \nu_k &= \int x^k \psi(x)\, dx. \end{aligned} \qquad (1.17)$$

They are related to the Fourier transforms of ϕ, ψ by

$$\begin{aligned} \mu_k &= \sqrt{2\pi}\, i^k D^k \hat{\phi}(0), \\ \nu_k &= \sqrt{2\pi}\, i^k D^k \hat{\psi}(0). \end{aligned} \qquad (1.18)$$

The continuous moment μ_0 is not determined by the refinement equation. It depends on the scaling of ϕ. For any given ϕ we can pick μ_0 arbitrarily, but for a biorthogonal pair the normalizations have to match (see lemma 1.25).

LEMMA 1.24
For all $x \in \mathbb{R}$,

$$\sum_k \phi(x-k) = \int \phi(x)\, dx = \mu_0.$$

PROOF From the basic regularity conditions,

$$\sum_k \phi(x-k) = c;$$

thus

$$c = \int_0^1 c\, dx = \sum_k \int_0^1 \phi(x-k)\, dx = \int_{-\infty}^\infty \phi(x)\, dx = \mu_0. \qquad \blacksquare$$

Chapter 1: Basic Theory

LEMMA 1.25
If $\phi, \tilde{\phi} \in L^1 \cap L^2$ are biorthogonal, then
$$\tilde{\mu}_0^* \mu_0 = 1.$$

PROOF   We expand the constant 1 in a series
$$1 = \sum_k \langle 1, \tilde{\phi}(x-k)\rangle \phi(x-k) = \tilde{\mu}_0^* \sum_k \phi(x-k) = \tilde{\mu}_0^* \mu_0.\quad\blacksquare$$

The normalization $\mu_0 = \tilde{\mu}_0 = 1$ is a natural choice. In the orthogonal case, the normalization $|\mu_0| = 1$ is required.

THEOREM 1.26
The continuous and discrete moments are related by
$$\mu_k = 2^{-k} \sum_{t=0}^{k} \binom{k}{t} m_{k-t} \mu_t,$$
$$\nu_k = 2^{-k} \sum_{t=0}^{k} \binom{k}{t} n_{k-t} \mu_t.$$
(1.19)

Once μ_0 has been chosen, all other continuous moments are uniquely defined and can be computed from these relations.

PROOF   This is proved in [71].
We start with
$$\hat{\phi}(2\xi) = h(\xi)\hat{\phi}(\xi)$$
and differentiate k times:
$$2^k \left(D^k \hat{\phi}\right)(2\xi) = \sum_{t=0}^{k} \binom{k}{t} D^{k-t} h(\xi) D^t \hat{\phi}(\xi).$$

When we set $\xi = 0$ and use equations (1.16) and (1.18), we get the first formula in equation (1.19). The second formula is proved similarly.
For $k = 0$, we get
$$\mu_0 = m_0 \mu_0,$$
which is always satisfied. μ_0 is an arbitrary nonzero number.
For $k \geq 1$, equation (1.19) leads to
$$(2^k - 1)\mu_k = \sum_{s=0}^{k-1} \binom{k}{s} m_{k-s} \mu_s.$$

We can compute $\mu_1, \mu_2, \ldots$ successively and uniquely from this. The second formula in equation (1.19) provides the ν_k. ∎

Example 1.7
For the scaling function of the Daubechies wavelet D_2, the first four discrete moments are
$$m_0 = 1, \quad m_1 = \frac{3-\sqrt{3}}{2}, \quad m_2 = \frac{6-3\sqrt{3}}{2}, \quad m_3 = \frac{27-17\sqrt{3}}{4}.$$
If we choose $\mu_0 = 1$, we can calculate
$$\mu_1 = \frac{3-\sqrt{3}}{2}, \quad \mu_2 = \frac{10-5\sqrt{3}}{2}, \quad \mu_3 = \frac{189-107\sqrt{3}}{28}. \qquad \Box$$

For later use, we note the following lemma.

LEMMA 1.27
Assume that μ_0 has been chosen to be 1. Then
$$n_k = 0, \quad k = 0, \ldots, p-1 \quad \Leftrightarrow \quad \nu_k = 0, \quad k = 0, \ldots, p-1$$
and
$$m_k = 0, \quad k = 1, \ldots, p-1 \quad \Leftrightarrow \quad \mu_k = 0, \quad k = 1, \ldots, p-1.$$

In other words, all continuous moments up to a certain order vanish if and only if all discrete moments up to that order vanish. (In the case of ϕ, the zeroth moment does not vanish, of course.)

The proof proceeds by induction, based on the formulas (1.19).

1.6 Approximation Order

As pointed out in section 1.2, the projection $P_n f$ of a function f onto the space V_n represents an approximation to f at resolution 2^{-n}. How good is this approximation?

DEFINITION 1.28 *The scaling function ϕ provides* approximation order p *if*
$$\|f - P_n f\| = O(2^{-np})$$

Chapter 1: Basic Theory

whenever f has p continuous derivatives.

If ϕ provides approximation order p, then for smooth f

$$\|Q_n f\| = O(2^{-np}),$$

since $\|Q_n f\| \leq \|f - P_n f\| + \|f - P_{n+1} f\|$.

DEFINITION 1.29 *The scaling function ϕ has accuracy p if all polynomials up to order $p-1$ can be represented as*

$$x^n = \sum_k c^*_{nk} \phi(x-k) \tag{1.20}$$

for some coefficients c_{nk}.

This representation is well-defined even though x^n does not lie in the space L^2. Since ϕ has compact support, the sum on the right is finite for any fixed x.

LEMMA 1.30
The coefficients c_{nk} in equation (1.20) have the form

$$c_{nk} = \sum_{t=0}^n \binom{n}{t} k^{n-t} y_t, \tag{1.21}$$

where $y_t = c_{t0}$, and $y_0 \neq 0$.

PROOF Replace x^n by $(x+\ell)^n$ in equation (1.20) and expand.
y_0 is $1/\mu_0$, which is nonzero. ∎

The y_k are called the *approximation coefficients*.
If ϕ has a dual $\tilde{\phi}$, we can multiply equation (1.20) by $\tilde{\phi}(x)$ and integrate to obtain

$$y_n^* = \langle x^n, \tilde{\phi}(x) \rangle = \tilde{\mu}_n^*. \tag{1.22}$$

DEFINITION 1.31 *The recursion coefficients $\{h_k\}$ of a refinement equation satisfy the* sum rules *of order p if*

$$\sum_k (-1)^k k^n h_k = 0$$

for $n = 0, \ldots, p-1$.

THEOREM 1.32

Assume ϕ satisfies the basic regularity conditions. Then the following are equivalent:

(i) ϕ has approximation order p.

(ii) ϕ has accuracy p.

(iii) $\{h_k\}$ satisfy the sum rules of order p.

(iv) The symbol h has a zero of order p at $\xi = \pi$, so it factors as

$$h(\xi) = \left(\frac{1 + e^{-i\xi}}{2}\right)^p h_0(\xi),$$

where h_0 is another trigonometric polynomial.

(v) For $k \in \mathbb{Z}$, $k \neq 0$, and $n = 0, \ldots, p-1$, $\hat{\phi}$ satisfies

$$\hat{\phi}(0) \neq 0,$$
$$D^n \hat{\phi}(2k\pi) = 0.$$

These are called the **Strang–Fix** conditions.

If $\phi(x)$ is part of a biorthogonal wavelet, the following are also equivalent to the above:

(vi) $\tilde{g}$ has a zero of order p at 0.

(vii) $\tilde{\psi}$ has p vanishing moments. That is, both the continuous and discrete moments of $\tilde{\psi}$ up to order $p-1$ are zero.

PROOF The full proof can be found in a number of places (e.g., in [87], [140]). We just give a sketch here.

(i) $\Leftrightarrow$ (ii): This is proved in [87]. It is the most technical part of the proof.

(ii) $\Rightarrow$ (iii): We already know that $\{h_k\}$ satisfy the sum rules of order 1 (theorem 1.10).

Assume that ϕ has approximation order p, and we have already established the sum rules of order n for some $n < p$. We expand

$$x^n = \sum_k c_{nk}^* \phi(x-k) = \sqrt{2} \sum_{k\ell} c_{nk}^* h_\ell \phi(2x - 2k - \ell)$$

and compare this to

$$(2x)^n = \sum_k c_{nk}^* \phi(2x - k).$$

Chapter 1: Basic Theory

Using the linear independence of translates of ϕ and the lower order sum rules, this gives us some relations for $\{h_k\}$. These can be reduced to the sum rule or order n. The details are lengthy and messy, unfortunately, so we will not present them here.

An alternative proof (also lengthy) can be found in [134].

(iii) $\Rightarrow$ (iv): This is quite easy:

$$(D^n h)(\pi) = (-i)^n \frac{1}{\sqrt{2}} \sum_k (-1)^k k^n h_k.$$

(iv) $\Rightarrow$ (v): For $\xi = 2k\pi$, $k = 2^n \ell$, ℓ odd, we apply equation (1.3) n times to get

$$\hat{\phi}(\xi) = h(2^{-n-1}\xi)\left[\prod_{s=1}^{n} h(2^{-s}\xi)\hat{\phi}(2^{-n-1}\xi)\right].$$

h has a zero of order p at $2^{-n-1}\xi = \ell\pi$ by periodicity, so $\hat{\phi}$ has a zero of order p at $\xi = 2k\pi$.

(v) $\Rightarrow$ (ii): The Poisson summation formula states

$$\sum_k \phi(x-k) = \sqrt{2\pi} \sum_k e^{2\pi i k x} \hat{\phi}(2\pi k),$$

so for $n \leq p-1$

$$\sum_k k^n \phi(x-k) = \sqrt{2\pi} i^n \sum_k D^n \left[e^{ix\xi}\hat{\phi}(\xi)\right]\bigg|_{\xi=2\pi k}.$$

The derivative on the right vanishes for all $k \neq 0$. The term for $k = 0$ is a polynomial in x of degree n, with nonzero leading coefficient. By induction, we can find polynomials $c_n(x)$ so that

$$\sum_k c_n(k)^* \phi(x-k) = x^n.$$

(iv) $\Leftrightarrow$ (vi): One of the biorthogonality relations in equation (1.12) is

$$g(\xi)\tilde{g}(\xi)^* + g(\xi+\pi)\tilde{g}(\xi+\pi)^* = 1.$$

If we assume that $\tilde{g}(0) = 0$, then $\tilde{g}(\pi) \neq 0$. The relation

$$h(\xi)\tilde{g}(\xi)^* + h(\xi+\pi)\tilde{g}(\xi+\pi)^* = 0 \tag{1.23}$$

for $\xi = 0$ then implies that $h(\pi) = 0$. Conversely, if $h(\pi) = 0$, then we must have $\tilde{g}(0) = 0$.

For higher p, we differentiate equation (1.23) repeatedly and set $\xi = 0$, with similar arguments.

(vi) ⇔ (vii): Condition (vi) is equivalent to vanishing discrete moments. The equivalence with vanishing continuous moments was lemma 1.27. ∎

In summary: approximation order p, accuracy p, and sum rules of order p are equivalent for sufficiently regular ϕ. They are also equivalent to the fact that the dual wavelet function has p vanishing moments (both discrete and continuous), and to a particular factorization of the symbol.

Example 1.8
We can now verify that the Daubechies wavelet D_2 has approximation order 2. The scaling function symbol factors as

$$h(\xi) = \left(\frac{1+e^{-i\xi}}{2}\right)^2 \left(\frac{1+\sqrt{3}}{2} + \frac{1-\sqrt{3}}{2}e^{-i\xi}\right).$$
☐

1.7 Symmetry

DEFINITION 1.33 *A function f is* symmetric *about the point a if*

$$f(a+x) = f(a-x) \quad \text{for all } x.$$

f is antisymmetric *about a if*

$$f(a+x) = -f(a-x) \quad \text{for all } x.$$

On the Fourier transform side, symmetry and antisymmetry are expressed as

$$\hat{f}(\xi) = \pm e^{-2ia\xi}\hat{f}(\xi).$$

$+$ corresponds to symmetry; $-$ corresponds to antisymmetry.

A scaling function cannot be antisymmetric, since antisymmetric functions automatically have integral zero. A scaling function of compact support can only be symmetric about the midpoint $a = (k_0+k_1)/2$ of its support, but such symmetric scaling functions exist (e.g., the Haar and hat scaling functions).

A wavelet function can be symmetric or antisymmetric.

LEMMA 1.34
A scaling function $\phi(x)$ of compact support which satisfies the refinement equation

$$\phi(x) = \sqrt{2} \sum_{k=k_0}^{k_1} h_k\, \phi(2x - k)$$

Chapter 1: Basic Theory

is symmetric about the point $a = (k_0 + k_1)/2$ if and only if the recursion coefficients are symmetric about this point:

$$h_{a+k} = h_{a-k},$$

or equivalently

$$h_{k_0+k} = h_{k_1-k}.$$

A wavelet function which satisfies the refinement equation

$$\psi(x) = \sqrt{2} \sum_{k=\ell_0}^{\ell_1} g_k\, \phi(2x - k)$$

has support $[(k_0 + \ell_0)/2, (k_1 + \ell_1)/2]$. If we let

$$b = \frac{\ell_0 + \ell_1}{2},$$

then ψ is symmetric or antisymmetric about the point

$$\frac{a+b}{2} = \frac{k_0 + \ell_0 + k_1 + \ell_1}{4}$$

if and only if the recursion coefficients are symmetric or antisymmetric about the point b:

$$g_{b+k} = g_{b-k}.$$

PROOF We calculate

$$\begin{aligned}\phi(a + x) &= \sqrt{2} \sum_k h_k \phi(2x + 2a - k) \\ &= \sqrt{2} \sum_k h_k \phi(a + [2x + a - k]) \\ &= \sqrt{2} \sum_k h_k \phi(a - [2x + a - k]) \\ &= \sqrt{2} \sum_k h_k \phi(-2x + k),\end{aligned}$$

while

$$\begin{aligned}\phi(a - x) &= \sqrt{2} \sum_k h_k \phi(-2x + 2a - k) \\ &= \sqrt{2} \sum_k h_{2a-k} \phi(-2x + k),\end{aligned}$$

so we get
$$h_k = h_{2a-k},$$
which is equivalent to the given conditions.

The second part is proved analogously. ∎

There are many examples of symmetric biorthogonal scaling functions, for example, the hat function. However, it is not possible for any *orthogonal* scaling function or wavelet function with compact support to be symmetric, except for the Haar wavelet.

THEOREM 1.35
If ϕ, ψ form an orthogonal wavelet with compact support, and at least one of ϕ or ψ is symmetric or antisymmetric, then it must be the Haar wavelet.

PROOF This is a sketch of the proof. The full proof can be found in [50, chapter 8].

If ψ is either symmetric or antisymmetric, the space W_k is invariant under the map $x \to (-x)$, so
$$V_k = \bigoplus_{\ell < k} W_\ell$$
is also invariant under this map.

V_0 then has two orthogonal basis functions: $\phi(x)$ and $\phi(-x)$. By lemma 1.13,
$$\phi(-x) = \alpha \phi(x + n)$$
for some α and $n \in \mathbb{N}$.

Both functions have the same integral, so $\alpha = 1$. Looking at the support, we find that n must be $k_0 + k_1$. This is an odd number, since an orthogonal scaling function has an even number of coefficients. ϕ is symmetric about the point $n/2$, so the recursion coefficients are symmetric:
$$h_k = h_{n-k}.$$

Then the orthogonality condition becomes
$$\delta_{0\ell} = \sum_k h_k h_{k-2\ell}^*$$
$$= \sum_k h_{2k} h_{2k-2\ell}^* + \sum_k h_{2k+1} h_{2k+1-2\ell}^*$$
$$= \sum_k h_{2k} h_{2k-2\ell}^* + \sum_k h_{n-2k-1} h_{n-2k-1+2\ell}^*$$
$$= 2 \sum_k h_{2k} h_{2k-2\ell}^*.$$

By lemma 1.12, this implies $h_{2k} = \beta\delta_{kN}$ for some $N \in \mathbb{N}$ and some β with $\beta^2 = 1/2$: precisely one of the even-numbered coefficients is nonzero. By symmetry, there is also precisely one nonzero odd-numbered coefficient.

It is then easy to check that this produces an orthogonal scaling function if and only if the nonzero coefficients are adjacent and both equal to $1/\sqrt{2}$. ϕ must be a (shifted) Haar wavelet. ∎

1.8 Point Values and Normalization

We mentioned the cascade algorithm earlier in this chapter as a practical way for finding approximate point values of $\phi(x)$. There is another approach which produces exact point values of ϕ. It usually works for continuous ϕ, but may fail in some cases.

For integer ℓ, $k_0 \leq \ell \leq k_1$, the refinement equation (1.1) reads

$$\phi(\ell) = \sqrt{2} \sum_{k=k_0}^{k_1} h_k\, \phi(2\ell - k) = \sqrt{2} \sum_{k=k_0}^{k_1} h_{2\ell-k}\, \phi(k).$$

This is an eigenvalue problem

$$\phi = T\phi, \tag{1.24}$$

where

$$\phi = \begin{pmatrix} \phi(k_0) \\ \phi(k_0+1) \\ \vdots \\ \phi(k_1) \end{pmatrix}, \qquad T_{\ell k} = \sqrt{2}\, h_{2\ell-k}, \quad k_0 \leq \ell, k \leq k_1.$$

Note that each column of T contains either all of the h_{2k} or all of the h_{2k+1}. The basic regularity condition

$$\sum_k h_{2k} = \sum_k h_{2k+1} = \frac{1}{\sqrt{2}}$$

implies that $(1, 1, \ldots, 1)$ is a left eigenvector to eigenvalue 1, so a right eigenvector also exists.

However, the eigenvalue 1 may be multiple: for the Haar wavelet, the matrix T is the 2×2 identity matrix. In some cases, a unique continuous solution can be found even if 1 is a multiple eigenvalue [19].

The first and last conditions in equation (1.24) are

$$\phi(k_0) = \sqrt{2}\, h_{k_0} \phi(k_0),$$
$$\phi(k_1) = \sqrt{2}\, h_{k_1} \phi(k_1).$$

Unless h_{k_0} or h_{k_1} are equal to $1/\sqrt{2}$, the values of ϕ at the endpoints are zero, and we can reduce the size of ϕ and T.

Once the values of ϕ at the integers have been determined, we can use the refinement equation to obtain values at the half-integers, then quarter-integers, and so on to any desired resolution.

Example 1.9
For the scaling function of the Daubechies wavelet D_2, the eigenvalue problem in equation (1.24) becomes

$$T\phi = \sqrt{2} \begin{pmatrix} h_0 & 0 & 0 & 0 \\ h_2 & h_1 & h_0 & 0 \\ 0 & h_3 & h_2 & h_1 \\ 0 & 0 & 0 & h_3 \end{pmatrix} \begin{pmatrix} \phi(0) \\ \phi(1) \\ \phi(2) \\ \phi(3) \end{pmatrix} = \begin{pmatrix} \phi(0) \\ \phi(1) \\ \phi(2) \\ \phi(3) \end{pmatrix}.$$

Since h_0, h_3 are not $1/\sqrt{2}$, we know $\phi(0) = \phi(3) = 0$, and we can reduce the problem to

$$\sqrt{2} \begin{pmatrix} h_1 & h_0 \\ h_3 & h_2 \end{pmatrix} \begin{pmatrix} \phi(1) \\ \phi(2) \end{pmatrix} = \begin{pmatrix} \phi(1) \\ \phi(2) \end{pmatrix}.$$

The solution, normalized to $\phi(1) + \phi(2) = 1$, is

$$\phi(1) = \frac{1 + \sqrt{3}}{2}, \quad \phi(2) = \frac{1 - \sqrt{3}}{2}.$$

Then

$$\phi(1/2) = \sqrt{2}\,[h_0 \phi(1) + h_1 \phi_0] = \frac{2 + \sqrt{3}}{4}. \quad □$$

We can also use the eigenvalue approach to compute point values of derivatives of ϕ (assuming they exist).

If ϕ satisfies the refinement equation

$$\phi(x) = \sqrt{2} \sum_{k=k_0}^{k_1} h_k\, \phi(2x - k),$$

then

$$(D\phi)(x) = 2\sqrt{2} \sum_{k=k_0}^{k_1} h_k\, (D\phi)(2x - k).$$

With the same derivation as above, we obtain

$$D\phi = 2T(D\phi),$$

so $D\phi$ is an eigenvector of T to eigenvalue $1/2$. This eigenvalue must exist if ϕ is continuously differentiable.

Chapter 1: Basic Theory 37

The nth derivative can be likewise computed from the eigenvector to eigenvalue 2^{-n}.

LEMMA 1.36
The correct normalization for the nth derivative is

$$\sum_k k^n D^n \phi(k) = (-1)^n n! \mu_0.$$

PROOF As in the proof of lemma 1.24, we can show that

$$\sum_k k^n D^n \phi(k) = \int x^n D^n \phi(x)\, dx.$$

After integrating by parts n times, this becomes

$$\sum_k k^n D^n \phi(k) = (-1)^n n! \int \phi(x)\, dx = (-1)^n n! \mu_0. \qquad \blacksquare$$

Example 1.10
Our standard example D_2 is not differentiable.

The scaling function of the Daubechies wavelet D_3 has recursion coefficients listed in section A. It is continuously differentiable.

With the normalizations

$$\sum_k \phi(k) = 1, \qquad \sum_k k\phi(k) = -1$$

we get

$$\begin{pmatrix} \phi(0) \\ \phi(1) \\ \phi(2) \\ \phi(3) \\ \phi(4) \\ \phi(5) \end{pmatrix} = \begin{pmatrix} 0 \\ 1.28633506942570 \\ -0.38583696104588 \\ 0.09526754600378 \\ 0.00423434561640 \\ 0 \end{pmatrix},$$

$$\begin{pmatrix} D\phi(0) \\ D\phi(1) \\ D\phi(2) \\ D\phi(3) \\ D\phi(4) \\ D\phi(5) \end{pmatrix} = \begin{pmatrix} 0 \\ 1.63845234088407 \\ -2.23275819046311 \\ 0.55015935827401 \\ 0.04414649130504 \\ 0 \end{pmatrix}. \qquad \square$$

2
Practical Computation

As explained in sections 1.2, 1.4, the discrete wavelet transform (DWT) is based on the decomposition

$$V_n = V_\ell \oplus W_\ell \oplus W_{\ell+1} \oplus \cdots \oplus W_{n-1}.$$

A function $s \in V_n$ can be expanded either as

$$s = \sum_k s_{nk}^* \phi_{nk}$$

or as

$$s = \sum_k s_{\ell k}^* \phi_{\ell k}(x) + \sum_{j=\ell}^{n-1} \sum_k d_{jk}^* \psi_{jk},$$

where

$$\phi_{nk}(x) = 2^{n/2} \phi(2^n x - k)$$

(and likewise for ψ, $\tilde\phi$, $\tilde\psi$), and

$$s_{nk}^* = \langle s, \tilde\phi_{nk} \rangle,$$
$$d_{nk}^* = \langle s, \tilde\psi_{nk} \rangle.$$

We put complex conjugates on the coefficients with a view toward the second part of the book, where these coefficients will be row vectors.

The notations s and d originally stood for *sum* and *difference*, which is what they are for the Haar wavelet. You can also think of them as standing for the *smooth* part and the fine *detail* of s. The original function $s(x)$ is the *signal*.

The DWT and inverse DWT (IDWT) convert the s_{nk} into $s_{\ell k}$, d_{jk}, $j = \ell, \ldots, n-1$, and conversely. The implementation is described in section 2.1.

The DWT algorithm requires the initial expansion coefficients s_{nk}. Frequently, the available data consist of equally spaced samples of s of the form $s(2^{-n}k)$. Converting $s(2^{-n}k)$ to s_{nk} is called *preprocessing*. After an IDWT, converting s_{nk} back to $s(2^{-n}k)$ is called *postprocessing*. Postprocessing has to be the inverse of preprocessing if we want to achieve perfect reconstruction.

In many applications of the DWT, the preprocessing step is skipped. We will discuss the validity of this approach in section 2.2, and mention some preprocessing approaches that have been proposed.

A very important question is the handling of boundaries. The DWT is defined for infinitely long signals. In practice, we can only handle finitely long signals. What do we do near the ends? Several approaches are covered in section 2.3.

We will then describe some alternative formulations of the DWT algorithm, and finally some methods for computing integrals involving scaling or wavelet functions.

2.1 Discrete Wavelet Transform

Assume that we have a function $s \in V_n$

$$s(x) = \sum_k s_{nk}^* \phi_{nk}(x),$$

represented by its coefficient vector $\mathbf{s}_n$. We decompose s into its components in V_{n-1}, W_{n-1}:

$$\begin{aligned}
s &= P_{n-1}s + Q_{n-1}s \\
&= \sum_j \langle s, \tilde{\phi}_{n-1,j}\rangle \phi_{n-1,j} + \sum_j \langle s, \tilde{\psi}_{n-1,j}\rangle \psi_{n-1,j} \\
&= \sum_j s_{n-1,j}^* \phi_{n-1,j} + \sum_j d_{n-1,j}^* \psi_{n-1,j}.
\end{aligned}$$

LEMMA 2.1

$$\begin{aligned}
\langle \phi_{n-1,j}, \tilde{\phi}_{nk}\rangle &= h_{k-2j}, \\
\langle \phi_{n-1,j}, \tilde{\psi}_{nk}\rangle &= g_{k-2j}, \\
\langle \tilde{\phi}_{n-1,j}, \phi_{nk}\rangle &= \tilde{h}_{k-2j}, \\
\langle \tilde{\psi}_{n-1,j}, \phi_{nk}\rangle &= \tilde{g}_{k-2j}.
\end{aligned}$$

PROOF We will just prove the first one:

$$\begin{aligned}
\langle \phi_{n-1,j}, \tilde{\phi}_{nk}\rangle &= \int 2^{(n-1)/2}\phi(2^{n-1}x - j) \cdot 2^{n/2}\tilde{\phi}(2^n x - k)\,dx \\
&= \int \sum_t 2^{n/2} h_t \phi(2^n x - 2j - t) \cdot 2^{n/2}\tilde{\phi}(2^n x - k)\,dx \\
&= \sum_t h_t \delta_{2j+t,k} = h_{k-2j}.
\end{aligned}$$ ∎

Chapter 2: Practical Computation 41

Using these formulas, we find that
$$s^*_{n-1,j} = \langle \sum_k s^*_{nk}\phi_{nk}, \tilde{\phi}_{n-1,j}\rangle = \sum_k s^*_{nk}\langle \phi_{nk}, \tilde{\phi}_{n-1,j}\rangle = \sum_k s^*_{nk}\tilde{h}^*_{k-2j},$$
or
$$s_{n-1,j} = \sum_k \tilde{h}_{k-2j} s_{nk}.$$

After similar calculations for $d_{n-1,j}$ and for the reconstruction step, we get the following algorithm.

ALGORITHM 2.2 Discrete Wavelet Transform (Direct Formulation)

The original signal is $\mathbf{s}_n = \{s_{nk}\}$.
Decomposition:
$$s_{n-1,j} = \sum_k \tilde{h}_{k-2j} s_{nk},$$
$$d_{n-1,j} = \sum_k \tilde{g}_{k-2j} s_{nk}.$$

The decomposed signal consists of two pieces $\mathbf{s}_{n-1}, \mathbf{d}_{n-1}$.
Reconstruction:
$$s_{nk} = \sum_j \left[h^*_{k-2j} s_{n-1,j} + g^*_{k-2j} d_{n-1,j} \right].$$

DEFINITION 2.3 *The* convolution $\mathbf{c} = \mathbf{a} * \mathbf{b}$ *of two sequences* $\mathbf{a}, \mathbf{b}$ *is defined by*
$$c_j = \sum_k a_{j-k} b_k.$$

If $\mathbf{a}$ is a sequence $\{\ldots, a_{-1}, a_0, a_1, \ldots\}$, we use the notation $\mathbf{a}(-)$ to denote the reversed sequence
$$\{\ldots, a_1, a_0, a_{-1}, \ldots\}.$$
The notation $(\downarrow 2)\mathbf{a}$ denotes the sequence *downsampled by 2*:
$$((\downarrow 2)\mathbf{a})_k = a_{2k}.$$
We throw the odd-numbered coefficients away and renumber the even-numbered coefficients.

The notation $(\uparrow 2)\mathbf{a}$ denotes the sequence *upsampled by 2*:
$$((\uparrow 2)\mathbf{a})_{2k} = a_k,$$
$$((\uparrow 2)\mathbf{a})_{2k+1} = 0.$$

We insert a zero between each pair of adjacent coefficients and renumber.

The decomposition step consists of two discrete convolutions

$$(\tilde{\boldsymbol{h}}(-) * \mathbf{s}_n)_j = \sum_k \tilde{h}_{-(j-k)} s_{nk},$$

$$(\tilde{\boldsymbol{g}}(-) * \mathbf{s}_n)_j = \sum_k \tilde{g}_{-(j-k)} s_{nk},$$

followed by downsampling:

$$\mathbf{s}_{n-1} = (\downarrow 2)(\tilde{\boldsymbol{h}}(-) * \mathbf{s}_n),$$
$$\mathbf{d}_{n-1} = (\downarrow 2)(\tilde{\boldsymbol{g}}(-) * \mathbf{s}_n).$$

The reconstruction step consists of upsampling, followed by a two convolutions:

$$\mathbf{s}_n = \mathbf{h}^* * (\uparrow 2)\mathbf{s}_{n-1} + \mathbf{g}^* * (\uparrow 2)\mathbf{d}_{n-1}.$$

What we have described so far is one step of the DWT. In practice, we do this over several levels:

$$\mathbf{s}_n \to \mathbf{s}_{n-1}, \mathbf{d}_{n-1}$$
$$\mathbf{s}_{n-1} \to \mathbf{s}_{n-2}, \mathbf{d}_{n-2}$$
$$\cdots$$
$$\mathbf{s}_{\ell+1} \to \mathbf{s}_\ell, \mathbf{d}_\ell.$$

The IDWT works similarly, in reverse.

How many floating point operations does this take?

If the original vector $\mathbf{s}_n$ has length N, it takes $O(N)$ operations for the first step. At the next step, $\mathbf{s}_{n-1}$ only has length $N/2$, so it takes $O(N/2)$ operations for the second step. Altogether, we have

$$O(N) + O(N/2) + O(N/4) + \cdots = O(N).$$

The fast Fourier transform (FFT) is implemented in a similar recursive manner. Each step turns a vector of length N into two new vectors half as long in $O(N)$ time. The difference is that in the FFT we need to continue working on *both* halves. That produces an operation count of

$$O(N) + O(N) + \cdots = O(N \log N).$$

The DWT is asymptotically faster than the Fourier transform, but this would only be noticeable for very long signals in practice.

The decomposition and reconstruction steps can also be interpreted as infinite matrix–vector products. The decomposition step is

$$\begin{pmatrix} \vdots \\ s_{n-1,-1} \\ s_{n-1,0} \\ s_{n-1,1} \\ \vdots \end{pmatrix} = \begin{pmatrix} \cdots & \cdots & \cdots & & & \\ \cdots & \tilde{h}_{-1} & \tilde{h}_0 & \tilde{h}_1 & \tilde{h}_2 & \cdots \\ & \cdots & \tilde{h}_{-1} & \tilde{h}_0 & \tilde{h}_1 & \tilde{h}_2 & \cdots \\ & & \cdots & \tilde{h}_{-1} & \tilde{h}_0 & \tilde{h}_1 & \tilde{h}_2 & \cdots \\ & & & \cdots & \cdots & \cdots & & \end{pmatrix} \begin{pmatrix} \vdots \\ s_{n,-1} \\ s_{n,0} \\ s_{n,1} \\ \vdots \end{pmatrix},$$

and similarly for the d-coefficients.

The matrix formulation becomes nicer if we interleave the s- and d-coefficients:

$$\begin{pmatrix} \vdots \\ s_{n-1,-1} \\ d_{n-1,-1} \\ s_{n-1,0} \\ d_{n-1,0} \\ s_{n-1,1} \\ d_{n-1,1} \\ \vdots \end{pmatrix} = \begin{pmatrix} \cdots & \cdots & \cdots & & & & & \\ \cdots & \tilde{h}_{-1} & \tilde{h}_0 & \tilde{h}_1 & \tilde{h}_2 & \cdots & & \\ \cdots & \tilde{g}_{-1} & \tilde{g}_0 & \tilde{g}_1 & \tilde{g}_2 & \cdots & & \\ & & \cdots & \tilde{h}_{-1} & \tilde{h}_0 & \tilde{h}_1 & \tilde{h}_2 & \cdots \\ & & \cdots & \tilde{g}_{-1} & \tilde{g}_0 & \tilde{g}_1 & \tilde{g}_2 & \cdots \\ & & & & \cdots & \tilde{h}_{-1} & \tilde{h}_0 & \tilde{h}_1 & \tilde{h}_2 & \cdots \\ & & & & \cdots & \tilde{g}_{-1} & \tilde{g}_0 & \tilde{g}_1 & \tilde{g}_2 & \cdots \\ & & & & \cdots & \cdots & \cdots & & \end{pmatrix} \begin{pmatrix} \vdots \\ s_{n,-1} \\ s_{n,0} \\ s_{n,1} \\ \vdots \end{pmatrix},$$

or simply

$$(\mathbf{sd})_{n-1} = \tilde{L}\,\mathbf{s}_n. \tag{2.1}$$

$\tilde{L}$ is an infinite banded block Toeplitz matrix

$$\tilde{L} = \begin{pmatrix} \cdots & \cdots & \cdots & \\ \cdots & \tilde{L}_0 & \tilde{L}_1 & \cdots \\ & \cdots & \tilde{L}_0 & \tilde{L}_1 & \cdots \\ & & \cdots & \cdots & \cdots \end{pmatrix}$$

with

$$\tilde{L}_k = \begin{pmatrix} \tilde{h}_{2k} & \tilde{h}_{2k+1} \\ \tilde{g}_{2k} & \tilde{g}_{2k+1} \end{pmatrix}.$$

The reconstruction step can be similarly written as

$$\mathbf{s}_n = L^*(\mathbf{sd})_{n-1}. \tag{2.2}$$

The perfect reconstruction condition is expressed as

$$L^*\tilde{L} = I.$$

Throughout this book, I stands for an identity matrix of appropriate size.

2.2 Pre- and Postprocessing

The DWT algorithm requires the initial expansion coefficients s_{nk}. Frequently, the available data consists of equally spaced samples of s of the form $s(2^{-n}k)$. Converting $s(2^{-n}k)$ to s_{nk} is called *preprocessing* or *prefiltering*. After an IDWT, converting s_{nk} back to $s(2^{-n}k)$ is called *postprocessing* or *postfiltering*. Postprocessing has to be the inverse of preprocessing if we want to achieve perfect reconstruction.

In [134, page 232], the authors call equating the expansion coefficients and point samples a "wavelet crime." I would not go quite so far, but a number of other authors also recommend a prefiltering step. For simplicity, we assume in this section that s is real.

Strang and Nguyen [134] recommend replacing the exact values

$$s_{nk} = \int s(x)\tilde{\phi}_{nk}(x)\,dx$$

by the trapezoidal rule sums

$$s_{nk} \approx 2^{-n/2} \sum_{\ell} s(2^{-n}\ell)\tilde{\phi}(\ell - k)\,dx.$$

Recall that the point values of $\tilde{\phi}$ at the integers are known.

Other suggested prefiltering algorithms can be found in [24, section 3.2] and [64]. If s is given in closed form and the coefficients s_{nk} are needed with high accuracy, they can be computed with the techniques of section 2.8.3.

Contrary to the above, I will now present a heuristic argument why it is often acceptable to use the expansion coefficients s_{nk} and function samples $s(2^{-n}k)$ interchangeably. We assume that $s(x)$ is given exactly and is at least twice differentiable. We also assume that $\tilde{\phi}$ is real-valued and has been normalized so that

$$\tilde{\mu}_0 = \int \tilde{\phi}(x)\,dx = 1.$$

Fix the level n. Then

$$s_{nk} = \langle s, \tilde{\phi}_{nk}\rangle = \int s(x)\,\tilde{\phi}_{nk}(x)\,dx$$
$$= \int s(x)\,2^{n/2}\tilde{\phi}(2^n x - k)\,dx \qquad (2.3)$$
$$= 2^{-n/2} \int s(2^{-n}(k+y))\tilde{\phi}(y)\,dy.$$

We expand s into a Taylor series about the point $2^{-n}(k + \tilde{\mu}_1)$, where $\tilde{\mu}_1$ is the first continuous moment of $\tilde{\phi}$:

$$s(2^{-n}(k+y)) = s(2^{-n}(k+\tilde{\mu}_1)) + s'(2^{-n}(k+\tilde{\mu}_1))2^{-n}(y - \tilde{\mu}_1) + O(2^{-2n}).$$

When we substitute this into equation (2.3), the derivative term vanishes, since

$$\int (y - \tilde{\mu}_1)\tilde{\phi}(y)\,dy = 0,$$

so

$$s_{nk} = 2^{-n/2}\left[s(2^{-n}(k + \tilde{\mu}_1)) + O(2^{-2n})\right] \approx 2^{-n/2}s(2^{-n}(k + \tilde{\mu}_1)).$$

Chapter 2: Practical Computation 45

For smooth s, the truncation error is smaller than the coefficients by a factor of order 2^{-2n}, which is quite small for larger n. Except for a scaling constant and a shift of $2^{-n}\tilde{\mu}_1$, the expansion coefficients and equally spaced samples are almost the same.

Postprocessing will provide point samples of the signal s. Point values in between the sampling points could be found by adding up the scaling function expansion, but this often gives the reconstructed signal a very ragged appearance. Many scaling functions are not very smooth.

Better alternatives for finding intermediate points include interpolation, or the approach described in [99].

2.3 Handling Boundaries

The DWT and IDWT, as described so far, operate on infinite sequences of coefficients. In real life, we can only work on finite sequences. How should we handle the boundary?

We want the finite length DWT to be linear, so it should be of the form

$$(\mathbf{sd})_{n-1} = \tilde{L}_n \mathbf{s}_n$$

for some matrix $\tilde{L}_n$, in analogy with equation (2.1). In order to preserve the usual definition of the DWT as much as possible, we postulate the form

$$\tilde{L}_n = \begin{pmatrix} \tilde{L}_b & & \\ & \tilde{L}_i & \\ & & \tilde{L}_e \end{pmatrix},$$

so that

- The interior part $\tilde{L}_i$ is a segment of the infinite block Toeplitz matrix $\tilde{L}$, and each row contains a complete set of coefficients. This part will make up most of the matrix. $\tilde{L}_i$ approximately doubles in size when n is increased by 1.

- The matrices $\tilde{L}_b$ at the beginning and $\tilde{L}_e$ at the end are fairly small and remain constant at all levels. These matrices will handle the boundaries.

- The entire matrix $\tilde{L}_n$ has the same block structure as $\tilde{L}$ (each block row shifted by one compared to its neighbors).

- $\tilde{L}_n$ is invertible, and its inverse matrix L_n^* has the analogous structure

$$L_n^* = \begin{pmatrix} L_b^* & & \\ & L_i^* & \\ & & L_e^* \end{pmatrix}.$$

In the orthogonal case, we also would like to preserve $L_n^{-1} = L_n^*$.

There are a number of ways to find suitable boundary coefficients. An excellent overview can be found in [134, section 8.5].

2.3.1 Data Extension Approach

This is very easy to implement. We artificially extend the signal across the boundaries so that each extended coefficient is a linear combination of known coefficients.

For example, suppose our signal starts with $s_{0,0}$ and we need to compute

$$s_{-1,0} = \tilde{h}_{-1} s_{0,-1} + \tilde{h}_0 s_{0,0} + \tilde{h}_1 s_{0,1} + \cdots.$$

If our extension method is

$$s_{0,-1} = \alpha s_{0,0} + \beta s_{0,1},$$

then

$$s_{-1,0} = (\tilde{h}_0 + \alpha \tilde{h}_{-1}) s_{0,0} + (\tilde{h}_1 + \beta \tilde{h}_{-1}) s_{0,1} + \cdots.$$

The h- and g-coefficients that "stick out over the side" of $\tilde{L}_n$ are wrapped back inside.

This gives us $\tilde{L}_n$, and its inverse will be L_n^*. There is no guarantee that $\tilde{L}_n$ will not be singular, or that $L_n = \tilde{L}_n^{-*}$ will have the correct form, but it usually works.

Specific data extension methods are:

Periodic Extension

The effect of this is that h- and g-coefficients that disappear on the left reappear on the right, and vice versa.

This always works and preserves orthogonality and approximation order 1, but it is not necessarily a good idea unless the data are truly periodic. The jump at the boundary leads to spurious large d-coefficients.

Zero Extension

We simply truncate the infinite matrix $\tilde{L}$. Zero padding introduces jumps, like periodic extension. It does not preserve orthogonality or any approximation orders.

Symmetric Extension

We reflect the data about the endpoints. The effect on $\tilde{L}$ is that coefficients that disappear at the end get reflected back. This is the recommended method for symmetric filters.

There are two ways of doing a reflection: if the data are

$$s_0, s_1, s_2, \ldots,$$

Chapter 2: Practical Computation

we can extend this by either

$$\ldots, s_2, s_1, s_0, s_1, s_2, \ldots$$

(whole-sample symmetry) or by

$$\ldots, s_2, s_1, s_0, s_0, s_1, s_2, \ldots$$

(half-sample symmetry). We can also do antisymmetric reflection, but in the whole-sample case this only works if $s_0 = 0$.

This approach is discussed in great detail in [27]. A finite signal can be extended to a periodic infinite signal through repeated reflections. For example, if the original signal is

$$s_0, s_1, s_2, s_3$$

and we use half-sample symmetry at both ends, this becomes

$$\ldots, s_1, s_0, s_0, s_1, s_2, s_3, s_3, s_2, s_1, s_0, s_0, s_1, \ldots$$

It is shown in [27] that if you match the type of data extension correctly to the type of symmetry of the scaling function, then a finite DWT based on symmetric extension will be equivalent to an infinite DWT on the symmetric extension: the DWT will preserve the symmetry across levels.

Constant or Linear Extrapolation

This is discussed in [150]. Constant extrapolation preserves approximation order 1; linear extrapolation preserves approximation order 2. They do not preserve orthogonality.

Example 2.1
For the Daubechies wavelet D_2, the decomposition matrix with linear extension looks like this:

$$\tilde{L}_n = \begin{pmatrix} h_1 + 2h_0 & h_2 - h_0 & h_3 & & & & & & \\ -h_2 + 2h_3 & h_1 - h_3 & -h_0 & & & & & & \\ & & h_0 & h_1 & h_2 & h_3 & & & \\ & & h_3 & -h_2 & h_1 & -h_0 & & & \\ & & & \ddots & \ddots & & & & \\ & & & & & h_0 & h_1 & h_2 & h_3 \\ & & & & & h_3 & -h_2 & h_1 & -h_0 \\ & & & & & & & h_0 & h_1 - h_3 & h_2 + 2h_3 \\ & & & & & & & h_3 & -h_2 + h_0 & h_1 - 2h_0 \end{pmatrix}.$$

The inverse has the correct structure, with slightly larger end blocks. □

2.3.2 Matrix Completion Approach

This is a linear algebra approach described in [108] and [134]. We look for suitable end blocks which guarantee

$$\tilde{L}_n L_n^* = I.$$

This can be done in a way that will preserve orthogonality, but it will not preserve approximation order in general.

Example 2.2
We take the Daubechies wavelet D_2. This wavelet is orthogonal, so we can drop the tildes. We look for L_n of the form

$$\begin{pmatrix} a & b & c & & & & & & \\ d & e & f & & & & & & \\ h_0 & h_1 & h_2 & h_3 & & & & & \\ h_3 & -h_2 & h_1 & -h_0 & & & & & \\ & & \ddots & \ddots & & & & & \\ & & & & h_0 & h_1 & h_2 & h_3 & \\ & & & & h_3 & -h_2 & h_1 & -h_0 & \\ & & & & & & u & v & w \\ & & & & & & x & y & z \end{pmatrix}. \quad (2.4)$$

The solution is not unique. Given any solution, we can multiply L_b by any 2×2 orthogonal matrix and get another solution; similarly, for L_e. Following the suggestion from [134], we impose the additional conditions

$$d + e + f = 0$$
$$x + y + z = 0.$$

This will ensure that constant vectors get annihilated by the wavelet coefficients. It does not give us approximation order 1, though: the scaling function coefficients do not preserve constant vectors.

We concentrate first on orthogonality: rows 1 and 2 have to be orthogonal to rows 3 and 4 (they are automatically orthogonal to the other rows). This is possible since (h_0, h_1) and $(h_3, -h_2)$ are linearly dependent. The vector normal to (h_0, h_1) and $(h_3, -h_2)$ is (h_2, h_3), so we start with

$$b = e = h_2$$
$$c = f = h_3$$
$$d = -e - f = -h_2 - h_3.$$

Orthogonality of the first two rows then requires

$$a = \frac{h_2^2 + h_3^2}{h_2 + h_3}.$$

Then we normalize each row. The result is

$$L_b = \begin{pmatrix} 0.93907080158804 & 0.29767351612730 & -0.17186188466672 \\ -0.34372376933344 & 0.81325917012746 & -0.46953540079402 \end{pmatrix}.$$

Similar calculations at the other end give

$$L_e = \begin{pmatrix} 0.40344911067754 & 0.69879435796197 & 0.59069049456887 \\ 0.29534524728444 & 0.51155297407064 & -0.80689822135507 \end{pmatrix}.$$

2.3.3 Boundary Function Approach

This approach is the most time-consuming, but it can preserve both orthogonality and approximation order.

The idea is to introduce special boundary functions at each end of the interval, and work out the resulting decomposition and reconstruction algorithms.

This approach was pioneered in [42], where it was carried out for the Daubechies wavelets. A more general description that can also incorporate boundary conditions can be found in [6].

The notation and explanation necessary for the general description is quite extensive. We will content ourselves with a detailed worked-out example.

Example 2.3

Consider again the Daubechies wavelet D_2. It has support length 3, so on an interval of length N there are $N-2$ integer translates of ϕ that are contained completely inside the interval. This suggests that we should use one boundary function at each end.

We only consider the left end, which we take to be 0. We assume the interval is sufficiently long so that there is no interference between the left and right boundary functions.

The interior functions are $\phi(x)$, $\phi(x-1)$, ... We look for a left boundary function ϕ_b that is a linear combination of the two scaling functions that cross the boundary, restricted to $x \geq 0$:

$$\phi_b(x) = \gamma \left[\alpha \phi(x+1) + \beta \phi(x+2) \right] \chi_{[0,\infty)}. \tag{2.5}$$

The parameter setup reflects the order of calculation: we first find α and β which make ϕ_b refinable. Then we adjust γ to make $\|\phi_b\| = 1$.

Since $\phi(x+1)$ and $\phi(x+2)$ are orthogonal to the interior functions, ϕ_b automatically inherits this property.

We want ϕ_b to be refinable, which means we want it to satisfy

$$\phi_b(x) = \sqrt{2} \left[a\phi_b(2x) + b\phi(2x) + c\phi(2x-1) \right]. \tag{2.6}$$

The range of functions to use in the refinement equation follows from considering the support: ϕ_b has support $[0, 2]$. The functions at level 1 that have support in $[0, 2]$ are $\phi_b(2x)$, $\phi(2x)$, $\phi(2x - 1)$.

Not every linear combination of the form in equation (2.5) is refinable. It is shown in [6] that if the scaling function has approximation order p, then projecting $x^0, x^1, \ldots, x^{q-1}$, $q \leq p$, onto the span of the boundary-crossing scaling functions will produce q refinable boundary functions that will preserve approximation order q.

We use $q = 1$. Since
$$\sum_k \phi(x - k) = 1,$$
the projection of the function 1 onto the span of $\phi(x + 1)$, $\phi(x + 2)$ is simply their sum: $\alpha = \beta = 1$.

Finding γ is a bit tricky, but with the methods described in section 2.8 we calculate
$$\gamma = \frac{1}{\sqrt{3} + 1}.$$

Then we work out the coefficients a, b, c from equation (2.6). They turn out to be $1/\sqrt{2}$, γh_2, and γh_3. Those numbers go directly into the first row of the decomposition matrix L_n in equation (2.4).

The boundary wavelet function ψ_b is found by looking for the orthogonal complement of V_0 in V_1. It is shown in [42] that it suffices if we take $\phi_{1b} \in V_1$ and subtract its component in direction ϕ_b:
$$\psi_b = \delta \left[\phi_{1b} - a\phi_{0b}(x)\right],$$
where
$$a = \langle \phi_{1b}, \phi_{0b} \rangle = 1/\sqrt{2}$$
is the same a as in equation (2.6). δ is chosen to ensure that $\|\psi_b\| = 1$.

In our example, we find $\delta = \sqrt{2}$, and
$$\psi_b = \frac{1}{\sqrt{2}}\phi_{1b} - \gamma h_2 \phi_{10} - \gamma h_3 \phi_{11}.$$

These numbers go into the second row of L_n.

We then need to repeat the calculations for the right end of the interval. Altogether we find that

$$\begin{pmatrix} 1/\sqrt{2} & \gamma_b h_2 & \gamma_b h_3 \\ 1/\sqrt{2} & -\gamma_b h_2 & -\gamma_b h_3 \\ & & h_0 & h_1 & h_2 & h_3 \\ & & h_3 & -h_2 & h_1 & -h_0 \\ & & & \ddots & \ddots \\ & & & & h_0 & h_1 & h_2 & h_3 \\ & & & & h_3 & -h_2 & h_1 & -h_0 \\ & & & & & & \gamma_e h_0 & \gamma_e h_1 & 1/\sqrt{2} \\ & & & & & & -\gamma_e h_0 & -\gamma_e h_1 & 1/\sqrt{2} \end{pmatrix},$$

Chapter 2: Practical Computation

where $\gamma_b = 1/(\sqrt{3}+1)$, $\gamma_e = \sqrt{3}-1$.
This matrix is orthogonal, and it preserves approximation order 1. □

The DWT matrix derived in the example only preserves approximation order 1. To preserve approximation order 2, we need two boundary functions at each end. To keep the total function count constant, we have to remove one interior function at each end. We build two boundary functions at the left end out of different linear combinations of $\phi(x+2)$, $\phi(x+1)$, and $\phi(x)$. The interior functions begin with $\phi(x-1)$.

The calculations get more messy, and we have to include extra orthogonalization steps among the boundary functions, but the basic idea is the same. Details are given in [42].

2.3.4 Further Comments

Boundary methods usually require some preprocessing near the boundary.

Example 2.4
This is a continuation of the previous example.
We can calculate that

$$\int \phi_b(x)\,dx = \frac{1}{\sqrt{3}+1},$$

$$\int \phi_e(x)\,dx = \frac{\sqrt{3}+1}{2},$$

so the expansion of the constant 1 in the scaling function basis is

$$1 = \frac{1}{\sqrt{3}+1}\phi_b + \sum_k \phi(x-k) + \frac{\sqrt{3}+1}{2}\phi_e.$$

It is vectors of the form

$$(\frac{1}{\sqrt{3}+1}, 1, 1, \ldots, 1, 1, \frac{\sqrt{3}+1}{2})^*$$

instead of $(1, 1, \ldots, 1)^*$ which get annihilated by the wavelet functions and preserved by the scaling functions.

We need to apply the preprocessing step of multiplying the first entry of the signal by $1/(\sqrt{3}+1)$ and the last entry by $(\sqrt{3}+1)/2$, and corresponding postprocessing if we do a reconstruction. □

Other approaches to the boundary problem can be found in [46] and [113]. The former paper pays close attention to the stability of the boundary functions.

A Mathematica algorithm to find the coefficients for the boundary function approach can be found in [5].

Which boundary handling method should you use? Periodic extension works well for periodic data, but not in general. Symmetric extension is recommended for symmetric wavelets. The boundary function approach works well if you can find or derive the coefficients.

If none of that applies, my preference is linear extension. It is easy to use and does not introduce artificial jumps in the data.

2.4 Putting It All Together

A complete DWT for a one-dimensional signal of finite length goes like this:

- Do preprocessing (optional).

- Decide how to handle the boundaries.

- Apply the algorithm.

If we start with a signal s_n of length N, the first step will produce two signals s_{n-1} and d_{n-1}, each of length $N/2$. They can be stored in the place previously occupied by s_n. Then we repeat the process with s_{n-1}, and so on.

The programming is easier if we put the s_k at the beginning of the vector. The output from the DWT routine after several steps is then

$$\begin{pmatrix} s_\ell \\ d_\ell \\ d_{\ell+1} \\ \vdots \\ d_{n-1} \end{pmatrix}.$$

We can then extract the components.

Example 2.5

Figure 2.1 shows the decomposition of a signal over three levels, using the Daubechies(7,9) biorthogonal wavelet. I have used periodic extension, but the boundary handling method is irrelevant in this example. The chosen signal has enough zeros near the ends.

The different levels are displayed with different scales on the horizontal axis. The original signal (top row) had length 1024. The other rows have lengths 128, 128, 256, 512 (top to bottom). Normalizing the horizontal scales ensures that features appear at corresponding locations at each level.

Chapter 2: Practical Computation   53

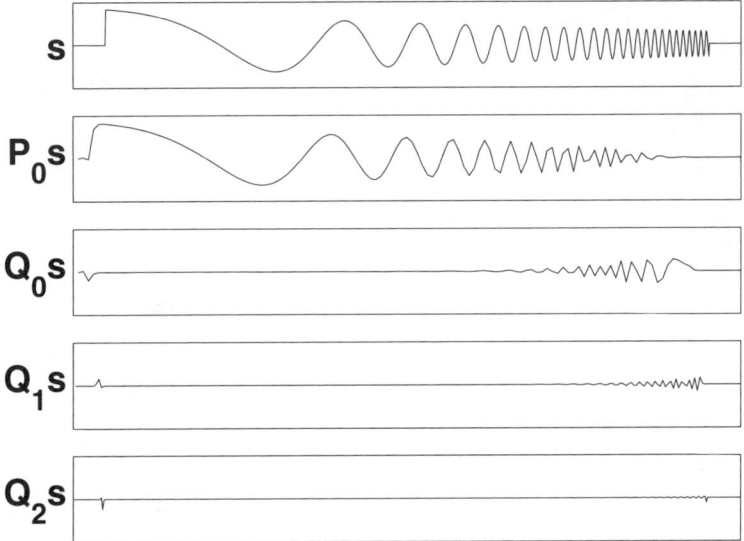

FIGURE 2.1
Decomposition of a signal over three levels.

I have also adjusted the vertical scale in each component. Each DWT step takes a constant vector into another constant vector that is half as long but multiplied by $\sqrt{2}$. The ℓ^2-norm is preserved. The numbers in row $P_0 s$ are actually larger than the numbers in row s by a factor of $\sqrt{8}$. □

Chapter 4 will discuss how to interpret the DWT. Right now we are just looking at the mechanics.

For two-dimensional signals (images), we do a DWT on both rows and columns.

After one level of decomposition, the image splits into four panels

$$S \to \begin{pmatrix} SS & DS \\ SD & DD \end{pmatrix}.$$

The first letter represents the filter that was applied in the horizontal direction; the second letter represents the vertical filter.

SS is a lower resolution version of the original. DS emphasizes vertical edges, since they correspond to jumps in the horizontal direction. SD emphasizes horizontal edges, and DD shows both.

For the second level of decomposition, we work on the top half in the vertical direction, and the left half in the horizontal direction. The image splits into nine pieces. Alternatively, we could just decompose SS further.

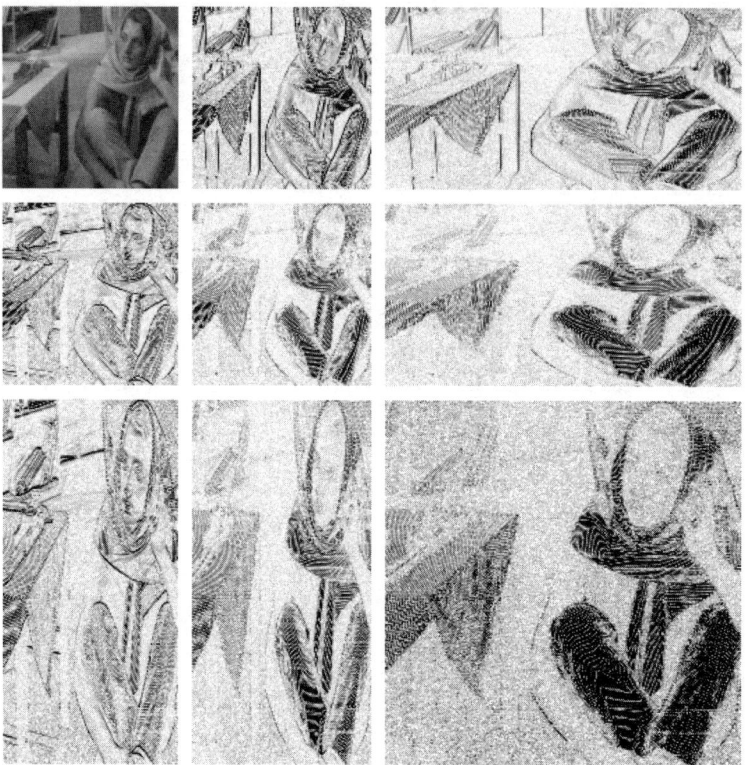

FIGURE 2.2
DWT of the Barbara image over two levels.

Example 2.6

Figure 2.2 shows the DWT of the Barbara image over two levels. A small version of the original image can be seen in the top left.

I have scaled each sub-image separately to make it visible. With uniform scaling, all except the top left subimage would be almost black. []

2.5 Modulation Formulation

The modulation formulation is a way of thinking about the DWT algorithm and verifying the perfect reconstruction conditions. It is not a way to actually implement it.

Chapter 2: Practical Computation   55

We associate with each sequence $\mathbf{a} = \{a_k\}$ (finite or infinite) its symbol

$$a(\xi) = \sum_k a_k e^{-ik\xi}.$$

If $\mathbf{c} = \mathbf{a} * \mathbf{b}$, then

$$c(\xi) = a(\xi)b(\xi).$$

Downsampling by two is represented as

$$(\downarrow 2)a(\xi) = \frac{1}{2}\left[a(\frac{\xi}{2}) + a(\frac{\xi}{2} + \pi)\right],$$

or

$$(\downarrow 2)a(2\xi) = \frac{1}{2}\left[a(\xi) + a(\xi + \pi)\right].$$

Upsampling by two is represented as

$$(\uparrow 2)a(\xi) = a(2\xi).$$

The entire DWT algorithm in terms of the symbols looks like this.

ALGORITHM 2.4 Discrete Wavelet Transform (Modulation Formulation)

The original signal is $s_n(\xi)$.
Decomposition:

$$s_{n-1}(2\xi) = \frac{1}{\sqrt{2}}\left[\tilde{h}(\xi)s_n(\xi) + \tilde{h}(\xi + \pi)s_n(\xi + \pi)\right]$$

$$d_{n-1}(2\xi) = \frac{1}{\sqrt{2}}\left[\tilde{g}(\xi)s_n(\xi) + \tilde{g}(\xi + \pi)s_n(\xi + \pi)\right].$$

The decomposed signal consists of two pieces $s_{n-1}(2\xi), d_{n-1}(2\xi)$.
Reconstruction:

$$\frac{1}{\sqrt{2}}s_n(\xi) = \left[h(\xi)^* s_{n-1}(2\xi) + g(\xi)^* d_{n-1}(2\xi)\right].$$

When we add the redundant statement

$$\frac{1}{\sqrt{2}}s_n(\xi + \pi) = \left[h(\xi + \pi)^* s_{n-1}(2\xi) + g(\xi + \pi)^* d_{n-1}(2\xi)\right]$$

to the reconstruction formula, we can write decomposition and reconstruction in the matrix form

$$\begin{pmatrix} s_{n-1}(2\xi) \\ d_{n-1}(2\xi) \end{pmatrix} = \begin{pmatrix} \tilde{h}(\xi) & \tilde{h}(\xi + \pi) \\ \tilde{g}(\xi) & \tilde{g}(\xi + \pi) \end{pmatrix} \cdot \frac{1}{\sqrt{2}} \begin{pmatrix} s_n(\xi) \\ s_n(\xi + \pi) \end{pmatrix},$$

$$\frac{1}{\sqrt{2}} \begin{pmatrix} s_n(\xi) \\ s_n(\xi + \pi) \end{pmatrix} = \begin{pmatrix} h(\xi)^* & g(\xi)^* \\ h(\xi + \pi)^* & g(\xi + \pi)^* \end{pmatrix} \begin{pmatrix} s_{n-1}(2\xi) \\ d_{n-1}(2\xi) \end{pmatrix}.$$

DEFINITION 2.5 *The matrix*

$$M(\xi) = \begin{pmatrix} h(\xi) & h(\xi+\pi) \\ g(\xi) & g(\xi+\pi) \end{pmatrix} \qquad (2.7)$$

is called the modulation matrix.

The biorthogonality conditions become

$$M(\xi)^* \tilde{M}(\xi) = I. \qquad (2.8)$$

DEFINITION 2.6 *A trigonometric polynomial matrix with the property*

$$U(\xi)^* U(\xi) = I \qquad (2.9)$$

is called paraunitary.

This generalizes the standard notion of a unitary matrix. The modulation matrix of an orthogonal refinable function is paraunitary.

REMARK 2.7 You may be wondering at this point where the factors of $1/\sqrt{2}$ in algorithm 2.4 come from.

One answer is that the symbols $h(\xi)$, $g(\xi)$ are defined differently from the symbols $s(\xi)$, $d(\xi)$: they carry a factor of $1/\sqrt{2}$ already. Together with the factor $1/\sqrt{2}$ in the decomposition formula, that makes up the factor of $1/2$ in the downsampling.

The matrix formulation makes it more clear why these factors have to be there. If ϕ, ψ form an orthogonal wavelet, the modulation matrix is paraunitary; it preserves two-norms:

$$\|s_n\|^2 = \|s_{n-1}\|^2 + \|d_{n-1}\|^2.$$

However, at level n we also have the redundant $s_n(\xi+\pi)$. The factor of $1/\sqrt{2}$ makes the norms of $(s_n(\xi), s_n(\xi+\pi))/\sqrt{2}$ and $s_n(\xi)$ equal. ∎

2.6 Polyphase Formulation

Convolutions can be implemented very efficiently on a computer, much faster than random multiplications and additions. There is even special hardware available. Unfortunately, a direct implementation of the DWT in terms of convolutions means that half the computed values get thrown away afterward.

Chapter 2: Practical Computation

There is a way to arrange the calculations in the algorithm in a form that uses convolutions with no wasted computations. This is called the *polyphase implementation*.

DEFINITION 2.8 The even and odd phases $\mathbf{a}_0$, $\mathbf{a}_1$ of a sequence $\mathbf{a} = \{a_k\}$ are defined by

$$a_{0,k} = a_{2k},$$
$$a_{1,k} = a_{2k+1}.$$

We split both the signal and the recursion coefficients into even and odd phases. Then

$$\begin{aligned}
s_{n-1,j} &= \sum_k \tilde{h}_{k-2j} s_{nk} \\
&= \sum_k \tilde{h}_{2k-2j} s_{n,2k} + \sum_k \tilde{h}_{2k+1-2j} s_{n,2k+1} \\
&= \sum_k \tilde{h}_{0,k-j} s_{n0,k} + \sum_k \tilde{h}_{1,k-j} s_{n1,k}.
\end{aligned}$$

This is now a sum of two convolutions:

$$\mathbf{s}_{n-1} = \tilde{\mathbf{h}}_0(-) * \mathbf{s}_{n,0} + \tilde{\mathbf{h}}_1(-) * \mathbf{s}_{n,1}.$$

The other parts of the DWT algorithm can be adapted similarly.

Decomposition begins with splitting the input into two phases. Each phase is filtered (convolved) with two different filters, and the results added. In the reconstruction step, we compute the even and odd phases separately, and finally recombine them. The number of floating-point operations is unchanged from the direct implementation.

DEFINITION 2.9 The *polyphase symbols* of a sequence $\mathbf{a} = \{a_k\}$ are given by

$$a_0(\xi) = \sum_k a_{0,k} e^{-ik\xi} = \sum_k a_{2k} e^{-ik\xi},$$
$$a_1(\xi) = \sum_k a_{1,k} e^{-ik\xi} = \sum_k a_{2k+1} e^{-ik\xi}.$$

In symbol notation, the polyphase DWT algorithm can be written as

Decomposition: $\begin{pmatrix} s_{n-1}(\xi) \\ d_{n-1}(\xi) \end{pmatrix} = \begin{pmatrix} \tilde{h}_0(\xi) & \tilde{h}_1(\xi) \\ \tilde{g}_0(\xi) & \tilde{g}_1(\xi) \end{pmatrix} \begin{pmatrix} s_{n,0}(\xi) \\ s_{n,1}(\xi) \end{pmatrix}.$

Reconstruction: $\begin{pmatrix} s_{n,0}(\xi) \\ s_{n,1}(\xi) \end{pmatrix} = \begin{pmatrix} h_0(\xi)^* & g_0(\xi)^* \\ h_1(\xi)^* & g_1(\xi)^* \end{pmatrix} \begin{pmatrix} s_{n-1}(\xi) \\ d_{n-1}(\xi) \end{pmatrix}.$

DEFINITION 2.10 *The matrix*

$$P(\xi) = \begin{pmatrix} h_0(\xi) & h_1(\xi) \\ g_0(\xi) & g_1(\xi) \end{pmatrix}$$

is called the polyphase matrix.

The biorthogonality conditions become

$$P(\xi)^* \tilde{P}(\xi) = I. \qquad (2.10)$$

The polyphase matrix of an orthogonal wavelet is paraunitary.

REMARK 2.11 Note that the polyphase symbols of the recursion coefficients

$$h_0(\xi) = \sum_k h_{2k} e^{-ik\xi}$$
$$h_1(\xi) = \sum_k h_{2k+1} e^{-ik\xi}$$

do *not* get a factor of $1/\sqrt{2}$ like the regular symbols. ∎

2.7 Lifting

There is a distinct difference between the modulation matrix and the polyphase matrix: the modulation matrix has a particular structure. All the information is already contained in the first column; the second column is simply the first column with shifted argument.

If we multiply the modulation matrices of different wavelets together, the result does not have any particular significance. Conversely, if we want to multiply a modulation matrix by some factor to create another modulation matrix, that factor has to have a particular structure.

A polyphase matrix, on the other hand, is unstructured. If P_1, $\tilde{P}_1$ and P_2, $\tilde{P}_2$ both satisfy

$$P_k(\xi)^* \tilde{P}_k(\xi) = I,$$

Chapter 2: Practical Computation

then so do $P = P_1 P_2$ and $\tilde{P} = \tilde{P}_1 \tilde{P}_2$.

This makes it possible to create new wavelets from existing ones by multiplying the polyphase matrix by some appropriate factor, and it opens the possibility of factoring a given polyphase matrix into elementary steps. One such factorization is described in this section.

For simplicity, we switch to the z-notation at this point, where

$$z = e^{-i\xi}.$$

This lets us work with polynomials rather than trigonometric polynomials.

Using the symbolic polyphase notation, we assume that the original signal $s_n(z)$ has already been decomposed into $s_{n-1}(z)$, $d_{n-1}(z)$. We now add an extra step.

DEFINITION 2.12 *A lifting step consists of an operation of the form*

$$s_{\text{new},n-1}(z) = s_{n-1}(z) + a(z)d_{n-1}(z)$$
$$d_{\text{new},n-1}(z) = d_{n-1},$$

or in matrix notation

$$\begin{pmatrix} s_{\text{new},n-1}(z) \\ d_{\text{new},n-1}(z) \end{pmatrix} = \begin{pmatrix} 1 & a(z) \\ 0 & 1 \end{pmatrix} \begin{pmatrix} s_{n-1}(z) \\ d_{n-1}(z) \end{pmatrix}.$$

A dual lifting step consists of an operation of the form

$$s_{\text{new},n-1}(z) = s_{n-1}(z),$$
$$d_{\text{new},n-1}(z) = d_{n-1} + b(z)s_{n-1}(z)$$

or in matrix notation

$$\begin{pmatrix} s_{\text{new},n-1}(z) \\ d_{\text{new},n-1}(z) \end{pmatrix} = \begin{pmatrix} 1 & 0 \\ b(z) & 1 \end{pmatrix} \begin{pmatrix} s_{n-1}(z) \\ d_{n-1}(z) \end{pmatrix}.$$

A decomposition step with some polyphase matrix $\tilde{P}$, followed by a lifting or dual lifting step, is equivalent to decomposition step with the single polyphase matrix

$$\tilde{P}_{\text{new}}(z) = \begin{pmatrix} 1 & a(z) \\ 0 & 1 \end{pmatrix} \tilde{P}(z)$$

or

$$\tilde{P}_{\text{new}}(z) = \begin{pmatrix} 1 & 0 \\ b(z) & 1 \end{pmatrix} \tilde{P}(z).$$

The inverses are easy to compute:

$$\begin{pmatrix} 1 & a(z) \\ 0 & 1 \end{pmatrix}^{-1} = \begin{pmatrix} 1 & -a(z) \\ 0 & 1 \end{pmatrix},$$

$$\begin{pmatrix} 1 & 0 \\ b(z) & 1 \end{pmatrix}^{-1} = \begin{pmatrix} 1 & 0 \\ -b(z) & 1 \end{pmatrix},$$

so

$$P_{\text{new}}(z) = \begin{pmatrix} 1 & 0 \\ -a(z)^* & 1 \end{pmatrix} P(z)$$

or

$$P_{\text{new}}(z) = \begin{pmatrix} 1 & -b(z)^* \\ 0 & 1 \end{pmatrix} P(z).$$

This means that during the reconstruction, we have to add an extra step

$$s_{\text{new},n-1}(z) = s_{n-1}(z) - a(z)d_{n-1}(z),$$
$$d_{\text{new},n-1}(z) = d_{n-1}$$

or

$$s_{\text{new},n-1}(z) = s_{n-1}(z),$$
$$d_{\text{new},n-1}(z) = d_{n-1} - b(z)s_{n-1}(z)$$

before the original reconstruction steps. Of course, we can also see that directly.

Swelden's idea [138], [139] was to build up DWT algorithms by adding lifting steps to a very simple initial wavelet. His favorite initial wavelet is the "lazy wavelet" with $P = \tilde{P} = I$, which is nothing but a splitting of the signal into its phases.

It can be shown that every possible wavelet pair of compact support can be constructed that way.

THEOREM 2.13
The polyphase matrix of any biorthogonal wavelet of compact support can be factored into a diagonal matrix and several pairs of lifting and dual lifting steps:

$$P(z) = \begin{pmatrix} h_0(z) & h_1(z) \\ g_0(z) & g_1(z) \end{pmatrix}$$
$$= \begin{pmatrix} d_1(z) & 0 \\ 0 & d_2(z) \end{pmatrix} \begin{pmatrix} 1 & 0 \\ b_k(z) & 1 \end{pmatrix} \begin{pmatrix} 1 & a_k(z) \\ 0 & 1 \end{pmatrix} \cdots \begin{pmatrix} 1 & 0 \\ b_1(z) & 1 \end{pmatrix} \begin{pmatrix} 1 & a_1(z) \\ 0 & 1 \end{pmatrix},$$

where the diagonal terms $d_k(z)$ are monomials.

Chapter 2: Practical Computation

PROOF This was proved in [53]. The proof is based on the Euclidean algorithm for determining the greatest common divisor of two polynomials. As a first step, we divide h_1 by h_0 to get

$$h_1(z) = h_0(z)a(z) + q(z),$$

so that

$$\begin{pmatrix} h_0(z) & h_1(z) \end{pmatrix} = \begin{pmatrix} h_0(z) & q(z) \end{pmatrix} \begin{pmatrix} 1 & a(z) \\ 0 & 1 \end{pmatrix},$$

where q has a lower degree than h_0. We then divide h_0 by q and continue in this manner until we reach z 0 in the top row.

At this point, the diagonal terms have to be monomials, since the determinant is a monomial. One more step will reduce the second row. ∎

The lifting factor decomposition is not unique. Polynomial division with remainder is unique for standard polynomials, but here we are dealing with Laurent polynomials. We could start with the highest powers and work downward, or start with the lowest powers and work upward. For example,

$$z + 2 = (z+1) \cdot 1 + 1$$
$$= (z+1) \cdot 2 - z.$$

Either way, the quotient has degree 1, and the remainder has degree 0 (as a Laurent polynomial).

For practical implementation, we are looking for a form that requires as few multiplications as possible, so we can relax the form of the factors a bit. We can allow the lifting factors to have monomials on the diagonal, or put the diagonal matrix on the right.

Example 2.7

For the Daubechies wavelet D_2, the polyphase matrix is

$$P(z) = \begin{pmatrix} h_0 + h_2 z & h_1 + h_3 z \\ h_3 + h_1 z & -h_2 + h_0 z \end{pmatrix}.$$

Each pair of outputs $s_{n-1,k}$, $d_{n-1,k}$ requires eight multiplications and six additions.

The matrix P factors as

$$\begin{pmatrix} 1 & -\sqrt{3} \\ 0 & 1 \end{pmatrix} \begin{pmatrix} 1 & 0 \\ \frac{\sqrt{3}}{4} + \frac{\sqrt{3}-2}{4}z & 1 \end{pmatrix} \begin{pmatrix} 1 & 1 \\ 0 & z \end{pmatrix} \begin{pmatrix} \frac{\sqrt{6}+\sqrt{2}}{2} & 0 \\ 0 & \frac{\sqrt{6}-\sqrt{2}}{2} \end{pmatrix}.$$

In this form, each pair of outputs requires only five multiplications and four additions. ∎

There is one additional advantage to the lifting formulation: we can add additional processing steps to the decomposition and reconstruction, such as rounding or other forms of quantization.

For example, we could replace

$$s_{\text{new},n-1}(z) = s_{n-1}(z) + a(z)d_{n-1}(z) \qquad \text{(decomposition)}$$
$$s_{n-1}(z) = s_{\text{new},n-1}(z) - a(z)d_{n-1}(z) \qquad \text{(reconstruction)}$$

by

$$s_{\text{new},n-1}(z) = s_{n-1}(z) + \lfloor a(z)d_{n-1}(z) \rfloor \qquad \text{(decomposition)}$$
$$s_{n-1}(z) = s_{\text{new},n-1}(z) - \lfloor a(z)d_{n-1}(z) \rfloor \qquad \text{(reconstruction)}$$

where $\lfloor x \rfloor$ is the largest integer less than or equal to x.

This is a nonlinear operation, but the reconstruction exactly undoes the effect of the decomposition. We can construct integer-to-integer transforms this way, for example.

2.8 Calculating Integrals

Scaling and wavelet functions are usually not known in closed form. Nevertheless, it is possible to compute many kinds of integrals exactly or approximately.

2.8.1 Integrals with Other Refinable Functions

Assume that $\phi_1(x)$ and $\phi_2(x)$ are two refinable functions

$$\phi_1(x) = \sqrt{2} \sum_{k=k_0}^{k_1} h_{1k} \phi_1(2x - k),$$

$$\phi_2(x) = \sqrt{2} \sum_{\ell=\ell_0}^{\ell_1} h_{2\ell} \phi_1(2x - \ell).$$

We define

$$f(y) = \int \phi_1(x) \phi_2(x - y) \, dx. \qquad (2.11)$$

When we substitute the refinement equations into the definition of f and sort things out, we find that

$$f(y) = \sqrt{2} \sum_{k=k_0-\ell_1}^{k_1-\ell_0} c_k f(2y - k),$$

Chapter 2: Practical Computation

where

$$c_k = \frac{1}{\sqrt{2}} \sum_\ell h_{1\ell} h_{2,\ell-k}.$$

The function f is again refinable. It is easy to check that

$$\sum_k c_{2k} = \sum_k c_{2k+1} = \frac{1}{\sqrt{2}}$$

if both the h_{1k}- and h_{2k}-coefficients satisfy this.

An integral of the form

$$\int \phi_1(x)\phi_2(x-n)\,dx$$

is then nothing but $f(n)$, and we already know how to find point values of refinable functions at the integers.

Since

$$\int f(y)\,dy = \left(\int \phi_1(x)\,dx \right) \left(\int \phi_2(x)\,dx \right),$$

the correct normalization is

$$\sum_k f(k) = \left(\sum_k \phi_1(k) \right) \left(\sum_k \phi_2(k) \right).$$

As a by-product, this is also the way to compute $\|\phi\|_2$ for biorthogonal wavelets. We take $\phi_1 = \phi$, $\phi_2 = \phi^*$, normalize the point values of f at the integers correctly, and then take $f(0)$.

Example 2.8

Take $\phi_1 = \phi_2 = \phi(x) =$ the hat function. Then the c-coefficients are

$$\{c_{-2}, c_{-1}, c_0, c_1, c_2\} = \frac{1}{8\sqrt{2}}\{1, 4, 6, 4, 1\}.$$

We assume that ϕ has been normalized to $\sum_k \phi(k) = 1$, which is the standard hat function with a peak of height 1.

The nonzero point values of f at the integers, correctly normalized, are

$$f(-1) = f(1) = 1/6, \qquad f(0) = 2/3.$$

This means that

$$\int \phi(x)\phi(x \pm 1)\,dx = 1/6,$$

$$\int |\phi(x)|^2\,dx = \|\phi\|_2^2 = 2/3,$$

which we can also verify directly. □

This approach works for any kind of convolution or correlation

$$f(y) = \int \phi_1(x)\phi_2(\pm x \pm y)\, dx,$$

with our without a complex conjugate on the second term. We just get slightly different formulas for the c_k.

There are many generalizations:

- **Integrals involving several refinable functions.** The counterpart to equation (2.11) for three functions is

$$f(y, z) = \int \phi_1(x)\phi_2(x - y)\phi_3(x - z)\, dx.$$

This is a function of two variables, but it is again refinable. After substituting the refinement equations, we obtain

$$f(y, z) = \sqrt{2} \sum_{jk\ell} h_j h_{j-k} h_{j-\ell} f(2y - k, 2z - \ell).$$

If the ϕ_j have compact support, so does f; we can find the point values of f on the integer grid $\mathbb{Z}^2$ by solving an eigenvalue problem.

- **Integrals involving refinable functions at different levels**, such as

$$\int \phi_1(x)\phi_2(2x - k)\, dx.$$

Here we would use the refinement equations to reduce all functions to a common level first.

- **Integrals involving scaling and wavelet functions.** Again, we would use the refinement equations to reduce this to an integral involving only scaling functions.

- **Integrals involving derivatives of refinable functions**, such as

$$\int \phi_1(x)\phi_2'(x - k)\, dx.$$

Here we would define $f(y)$ as above and look for point values of f' instead.

- **Integrals over subintervals.** Here we add a Haar function, as in

$$\int_k^{k+1} \phi(x)\, dx = \int \phi(x)\phi_2(x - k)\, dx$$

with $\phi_2 =$ Haar function.

Chapter 2: Practical Computation 65

More details can be found in [20] and [48]. Kunoth wrote a program package for performing these calculations [100]. See appendix C for details.

Example 2.9
Let ϕ_1 be the scaling function of the Daubechies wavelet D_2, and $\phi_2 = \chi_{[0,1]}$ the Haar scaling function.

The c-coefficients are

$$\{c_{-1}, c_0, c_1, c_2, c_3\} = \frac{1}{8\sqrt{2}}\{1+\sqrt{3}, 4+2\sqrt{3}, 6, 4-2\sqrt{3}, 1-\sqrt{3}\}.$$

The nonzero point values of f at the integers, correctly normalized, are

$$f(0) = \int_0^1 \phi_1(x)\,dx = \frac{5+3\sqrt{3}}{12}$$

$$f(1) = \int_1^2 \phi_1(x)\,dx = \frac{1}{6}$$

$$f(2) = \int_2^3 \phi_1(x)\,dx = \frac{5-3\sqrt{3}}{12}.$$

□

2.8.2 Integrals with Polynomials

We already know how to calculate an integral of the form

$$\int p(x)\phi(x)\,dx$$

where p is a polynomial. This is nothing but a linear combination of the continuous moments defined in equation (1.17)

$$\mu_k = \int x^k \phi(x)\,dx, \qquad (2.12)$$

which can be calculated from equation (1.19).

2.8.3 Integrals with General Functions

Here we want to approximate integrals of the form

$$s_{0k} = \int s(x)\phi(x-k)\,dx \qquad (2.13)$$

where s is a general function. This is useful, for example, for generating the true expansion coefficients of s (see section 2.2), or for Galerkin methods based on scaling functions as the basis functions.

The idea is to reduce such an integral to one of the cases that we already know: approximate s by a polynomial, or approximate s by a scaling function expansion. In either approach, we want to start at a sufficiently fine level n by calculating

$$s^*_{nk} = \int s(x) 2^{n/2} \phi(2^n x - k) \, dx. \tag{2.14}$$

To estimate this integral, we only have to approximate s on the support of ϕ_{nk}, which is a very small interval. Getting from the s_{nk} to s_{0k} is simply a DWT through n levels, with $d_{\ell k} = 0$ at all levels.

The polynomial or **quadrature approach** is described in [140]. It begins by deriving a quadrature formula

$$\int s(x) \phi(x) \, dx \approx \sum_\ell w_\ell s(x_\ell)$$

following the usual steps. Choose q quadrature points x_ℓ, write out the Lagrange interpolating polynomials at these points, and integrate to obtain the weights.

At level n, this rule has accuracy $O(h^q)$, $h = 2^{-n}$. What is interesting is that after going back to level 0, the integral still has accuracy $O(h^q)$. This is different from the usual repeated quadrature rules, where the global error has one order of accuracy less than the local error on each panel. The difference is that we are not adding the panels; we are applying a DWT.

In principle we could put the quadrature points wherever we want (e.g., equally spaced across the support of ϕ), but for efficiency we want some of the points to overlap when we shift by integers. We would also like to spread them out across the support of ϕ to get better accuracy.

Sweldens and Piessens [140] recommend to choose a starting point τ somewhere near the left end of the support of ϕ, and to choose a step size h which is a power of 2 so that the points

$$x_\ell = \tau + \ell h, \qquad \ell = 0, \ldots, q \tag{2.15}$$

are spread out as far as possible while remaining inside supp ϕ.

Example 2.10
We want to construct a three-point quadrature formula for the scaling function of the Daubechies wavelet D_2. The most regularly spaced points of the form in equation (2.15) are

$$x_0 = 1/2, \quad x_1 = 3/2, \quad x_2 = 5/2$$

for $\tau = 1/2$ and $h = 1$.

The continuous moments of ϕ are given in example 1.7.

The first quadrature weight is found by integrating the first Lagrange interpolating polynomial

$$L_0(x) = \frac{(x - 3/2)(x - 5/2)}{(1/2 - 3/2)(1/2 - 5/2)} = \frac{x^2 - 4x + 15/4}{2}.$$

It is
$$w_0 = \frac{\mu_2 - 4\mu_1 + (15/4)\mu_0}{2} = \frac{3 + 2\sqrt{3}}{8}.$$
The other two weights are found to be
$$w_1 = \frac{1}{4}, \quad w_2 = \frac{3 - 2\sqrt{3}}{8}.$$

To get the best accuracy, we let τ be arbitrary, and go through the same calculations again. The resulting weights w_k will be quadratic polynomials in τ. For example,
$$w_0 = \frac{1}{2}\tau^2 + \frac{1}{2}\sqrt{3}\tau + \frac{1}{4}.$$
For any choice of τ, the resulting quadrature rule will integrate quadratic polynomials correctly. Cubic polynomials will be integrated correctly if
$$\left(w_0 x_0^2 + w_1 x_1^3 - w_2 x_2^3\right) - \mu_3 = 0.$$

This leads to a cubic equation in τ. It has three real solutions, but the only one inside the support of ϕ is $\tau = 0.56517923327320$. The corresponding weights are
$$w_0 = 0.89917335656764$$
$$w_1 = 0.13285792392236$$
$$w_2 = -0.03203128049000 \qquad \square$$

A quadrature approach based on Gauss-type formulas is described in [16] and [17].

The **scaling function approach** is described in detail in [20] and [48]. We expand s into a scaling function series
$$s(x) \approx \sum_k s_{nk} \phi_{2,nk}(x),$$
where the scaling function ϕ_2 could be different from ϕ. The coefficients s_{nk} are approximated by point values
$$s_{nk} \approx s(2^{-n}(x + \tau - k)),$$
where τ is the first continuous moment of ϕ_2. ϕ_2 can be chosen to make this approximation very accurate. The integral in equation (2.13) can then be computed by adding up integrals between ϕ and ϕ_2.

It is shown in [48] that for continuous f this approach converges uniformly on compact sets to the correct integral.

3
Creating Wavelets

We have seen some examples of scaling and wavelet functions in earlier chapters. In this chapter we will discuss general ways of modifying existing wavelets, or creating wavelets with desired properties from scratch. We will then derive some of the standard types.

Before we do that, we will discuss the two *completion problems*.

- Given ϕ, $\tilde{\phi}$, find the wavelet functions.

- Given ϕ, find $\tilde{\phi}$.

The first completion problem is easier, and has an essentially unique answer. The second completion problem has multiple answers.

3.1 Completion Problem
3.1.1 Finding Wavelet Functions

We assume that ϕ, $\tilde{\phi}$ are given biorthogonal scaling functions. We want to find the wavelet functions.

Recall the definition of the modulation matrix

$$M(\xi) = \begin{pmatrix} h(\xi) & h(\xi+\pi) \\ g(\xi) & g(\xi+\pi) \end{pmatrix}.$$

The biorthogonality conditions in equation (1.12) are equivalent to

$$M(\xi)^* \tilde{M}(\xi) = M(\xi)\tilde{M}(\xi)^* = I. \tag{3.1}$$

Let

$$\Delta(\xi) = \det M(\xi) = h(\xi)g(\xi+\pi) - h(\xi+\pi)g(\xi). \tag{3.2}$$

Then

$$\tilde{M}(\xi)^* = \begin{pmatrix} \tilde{h}(\xi)^* & \tilde{g}(\xi)^* \\ \tilde{h}(\xi+\pi)^* & \tilde{g}(\xi+\pi)^* \end{pmatrix} \tag{3.3}$$

and also

$$\tilde{M}(\xi)^* = M(\xi)^{-1} = \frac{1}{\Delta(\xi)} \begin{pmatrix} g(\xi+\pi) & -h(\xi+\pi) \\ -g(\xi) & h(\xi) \end{pmatrix} \tag{3.4}$$

by using the well-known formula for the inverse of a 2 × 2 matrix.

By assumption, the entries in $\tilde{M}(\xi)$ are trigonometric polynomials of finite degree. This is only possible if $\Delta(\xi)$ is a monomial. Since $\Delta(\xi + \pi) = -\Delta(\xi)$ by equation (3.2), this monomial must be of odd degree:

$$\Delta(\xi) = \alpha e^{i(2n+1)\xi}, \qquad \alpha \neq 0, \quad n \in \mathbb{Z}.$$

In the orthogonal case, $M(\xi)$ is paraunitary, and α must satisfy $|\alpha| = 1$. In the real case, $\alpha = \pm 1$.

Equations (3.3) and (3.4) provide the necessary relations between the recursion coefficients. By comparing entries, we find that

$$\tilde{g}(\xi) = -\frac{1}{\Delta(\xi)^*} h(\xi + \pi)^*,$$

or in terms of the recursion coefficients

$$\tilde{g}_k = -\frac{1}{\alpha^*}(-1)^k h^*_{2n+1-k}. \tag{3.5}$$

Likewise,

$$g_k = -\alpha(-1)^k \tilde{h}^*_{2n+1-k}. \tag{3.6}$$

With these choices, it can be verified that the biorthogonality relation for h, $\tilde{h}$

$$h(\xi)\tilde{h}(\xi)^* + h(\xi + \pi)\tilde{h}(\xi + \pi)^* = 1, \tag{3.7}$$

implies all other biorthogonality relations.

3.1.2 Finding Dual Scaling Functions

Given $h(\xi)$, we want to find $\tilde{h}(\xi)$ so that

$$h(\xi)\tilde{h}(\xi)^* + h(\xi + \pi)\tilde{h}(\xi + \pi)^* = 1. \tag{3.8}$$

This can be converted into something called a *Bezout equation*, which we will study in more detail in a few pages. It has a solution if and only if $h(\xi)$ and $h(\xi + \pi)$ have no common zeros.

There is also the basic linear algebra approach. Equation (3.8) is a linear system of equations for $\tilde{h}_k$. Given ϕ, we choose the support of $\tilde{\phi}$, making sure it overlaps the support of ϕ. We convert equation (3.8) to a system of linear equations and try to solve it. Bezout's theorem (theorem 3.1) proves that a solution always exists if we take a sufficiently long support for $\tilde{\phi}$ (again assuming that $h(\xi)$ and $h(\xi + \pi)$ have no common zeros).

Additional constraints, such as symmetry or vanishing moments, can be added as extra equations.

Example 3.1
The Haar function has recursion coefficients

$$\{h_0, h_1\} = \{\frac{1}{\sqrt{2}}, \frac{1}{\sqrt{2}}\}.$$

Suppose we want to find a dual with coefficients $\tilde{h}_0, \ldots, \tilde{h}_3$. The biorthogonality conditions are

$$\tilde{h}_0 + \tilde{h}_1 = \sqrt{2}/2$$
$$\tilde{h}_2 + \tilde{h}_3 = 0.$$

That leaves two degrees of freedom. We could impose the sum rules of order 0 and 1

$$\tilde{h}_0 - \tilde{h}_1 + \tilde{h}_2 - \tilde{h}_3 = 0$$
$$-\tilde{h}_1 + 2\tilde{h}_2 - 3\tilde{h}_3 = 0$$

to get a dual with approximation order 2

$$\{\tilde{h}_0, \tilde{h}_1, \tilde{h}_2, \tilde{h}_3\} = \{\frac{3}{4\sqrt{2}}, \frac{5}{4\sqrt{2}}, \frac{1}{4\sqrt{2}}, -\frac{1}{4\sqrt{2}}\}.$$

These coefficients are similar in magnitude to those of the scaling function of the Daubechies wavelet D_2. The graph of the two functions are also quite similar.

We cannot impose symmetry with a dual of length 4, but there exists a symmetric dual scaling function of length 6. □

3.2 Projection Factors

This section is needed as background material later in this chapter. We just state some results without proof at this point. This material will be presented in greater generality in chapter 9.

It can be shown that the polyphase matrix of every orthogonal wavelet can be factored in the form

$$P(z) = QF_1(z)F_2(z)\cdots F_n(z)z^k, \qquad (3.9)$$

where Q is a constant orthogonal matrix and each F_k is a *projection factor* of the form

$$F(z) = (I - \mathbf{uu}^*) + \mathbf{uu}^*z$$

for some unit vector
$$\mathbf{u} = \begin{pmatrix} \cos\theta \\ \sin\theta \end{pmatrix}.$$

Projection factors are paraunitary. The number of factors is the *McMillan degree* of P, which is also the polynomial degree of $\det P(z)$. The factor z^k at the end only shifts the support of ϕ and is otherwise unimportant.

Example 3.2
The polyphase matrix of the Daubechies wavelet D_2 factors as

$$P(z) = \frac{1}{\sqrt{2}} \begin{pmatrix} 1 & 1 \\ 1 & -1 \end{pmatrix} \left[\frac{1}{4} \begin{pmatrix} 1 & \sqrt{3} \\ \sqrt{3} & 3 \end{pmatrix} + \frac{1}{4} \begin{pmatrix} 3 & -\sqrt{3} \\ -\sqrt{3} & 1 \end{pmatrix} z \right].$$

The second term is a projection factor based on the vector

$$\mathbf{u} = \frac{1}{2} \begin{pmatrix} \sqrt{3} \\ -1 \end{pmatrix}.$$
□

A popular alternative way to write projection factors is to use the parameter $\nu = \tan\theta$ instead of θ. This works out to

$$F(z) = \frac{1}{1+\nu^2} \begin{pmatrix} \nu^2 + z & \nu(z-1) \\ \nu(z-1) & 1+\nu^2 z \end{pmatrix}. \tag{3.10}$$

It can be shown (section 9.1.1) that the scaling function has approximation order 1 (or higher) if and only if

$$Q = \frac{1}{\sqrt{2}} \begin{pmatrix} 1 & 1 \\ \pm 1 & \mp 1 \end{pmatrix}. \tag{3.11}$$

There are also biorthogonal projection factors of the form

$$F(z) = (I - \mathbf{u}\mathbf{v}^*) + \mathbf{u}\mathbf{v}^* z$$

with $\mathbf{u}^*\mathbf{v} = 1$. The dual is

$$\tilde{F}(z) = (I - \mathbf{v}\mathbf{u}^*) + \mathbf{v}\mathbf{u}^* z.$$

It is *not* true, however, that every biorthogonal wavelet can be factored into such pieces. This will be discussed in more detail in chapter 9.

3.3 Techniques for Modifying Wavelets

A technique for modifying wavelets is one which creates new wavelets ϕ_{new}, ψ_{new}, $\tilde{\phi}_{\text{new}}$, $\tilde{\psi}_{\text{new}}$ by applying some transformation to given ϕ, ψ, $\tilde{\phi}$, $\tilde{\psi}$. Some examples are

Chapter 3: Creating Wavelets

- **Shifting factors between h, $\tilde{h}$.** If h, $\tilde{h}$ satisfy the biorthogonality relations in equation (3.7), and h factors in some manner

$$h(\xi) = f(\xi) h_0(\xi)$$

with $h_0(0) = 1$, then

$$h_{\text{new}}(\xi) = h_0(\xi),$$
$$\tilde{h}_{\text{new}}(\xi) = f(\xi)^* \tilde{h}(\xi)$$

are again biorthogonal. Examples can be found in sections 3.8 and 3.9.

- **Using projection factors.** We can add projection factors to an existing wavelet. This method has the advantage that it preserves orthogonality if F is orthogonal. Unfortunately, it also destroys approximation order beyond $p = 1$.

- **Using lifting factors.** We can add lifting factors or dual lifting factors to an existing wavelet. Appropriately chosen lifting factors can preserve approximation order, or even increase it, but they destroy orthogonality.

3.4 Techniques for Building Wavelets

A technique for building wavelets is one which creates wavelets from scratch. Some examples are

- **Solving the biorthogonality relation directly.** We look for trigonometric polynomial solutions of

$$h(\xi)\tilde{h}(\xi)^* + h(\xi + \pi)\tilde{h}(\xi + \pi)^* = 1.$$

This is equivalent to a system of quadratic equations. It can be solved directly for short wavelets, but that quickly becomes unwieldy.

Daubechies has found better ways to solve the orthogonality relation, based on the Bezout equation. We will see the details later in this chapter.

- **Using projection factors.** This is an easy way to create orthogonal wavelets with approximation order 1 of any size. Higher approximation orders can be imposed as extra relations among the parameters. For short wavelets, this can be done in closed form. For longer ones, it can be done numerically.

Examples are presented in sections 3.6.2 and 3.7.2.

- **Using lifting factors.** This is an easy way to create biorthogonal wavelets of any size. Approximation order is easy to enforce (details will be presented in the multiwavelet context in the second part of the book), and symmetry can also be produced this way.

  Examples can be found in [11], [138], and [139].

3.5 Bezout Equation

This is necessary background material for the rest of this chapter.

THEOREM 3.1
(Bezout) If p_1, p_2 are polynomials of degree n_1, n_2, respectively, with no common zeros, then there exist unique polynomials q_1, q_2 of degree $n_2 - 1$, $n_1 - 1$, respectively, so that

$$p_1(y)q_1(y) + p_2(y)q_2(y) = 1. \tag{3.12}$$

The proof is given in [50]. It is constructive, based on the Euclidean algorithm for finding the greatest common divisor.

The theorem only says that the solutions of lowest degree are unique. There may be other solutions of higher degree.

In the special case $p_1(y) = (1-y)^n$, $p_2(y) = \pm y^n$, we can find the solution explicitly. Solving equation (3.12) for q_1 gives

$$\begin{aligned} q_1(y) &= p_1(y)^{-1}\left[1 - p_2(y)q_2(y)\right] \\ &= (1-y)^{-n}\left[1 \mp y^n q_2(y)\right]. \end{aligned} \tag{3.13}$$

The shortest solution q_1 has degree $n-1$; it has to match the Taylor series expansion of $(1-y)^{-n}$ up to the y^{n-1}-term. Since

$$D^k\left[(1-y)^{-n}\right] = n(n+1)\cdots(n+k-1)(1-x)^{-n-k},$$

the coefficient of y^k in the Taylor expansion of $(1-y)^{-n}$ is

$$\frac{n(n+1)\cdots(n+k-1)}{k!} = \frac{(n+k-1)!}{k!(n-1)!} = \binom{n+k-1}{k}.$$

The shortest solution of equation (3.12) is

$$q_n(y) = \sum_{k=0}^{n-1} \binom{n+k-1}{k} y^k.$$

Bezout's theorem guarantees that this is actually a solution. This can also be verified directly.

The same calculation for q_2 produces the corresponding result

$$q_2(y) = \pm q_n(y).$$

Equation (3.13) shows that the general solution has to have the form

$$\begin{aligned} q_1(y) &= q_n(y) + y^n r_1(y), \\ q_2(y) &= \pm q_n(y) + y^n r_2(y). \end{aligned} \qquad (3.14)$$

It is easy to verify directly that this solves the Bezout equation if and only if

$$r_1(y) + r_2(1-y) = 0.$$

LEMMA 3.2
The general solution of the Bezout equation

$$(1-y)^n q_1(y) \pm y^n q_2(1-y) = 1$$

is

$$\begin{aligned} q_1(y) &= \sum_{k=0}^{n-1} \binom{n+k-1}{k} y^k + y^n r(y), \\ q_2(y) &= \pm \sum_{k=0}^{n-1} \binom{n+k-1}{k} y^k - y^n r(1-y), \end{aligned} \qquad (3.15)$$

where r is an arbitrary polynomial.

One observation that will be used later is that it is not actually necessary that r is a polynomial: any function will do. Of course, in that case q_1 and q_2 will be nonpolynomial solutions of the Bezout equation.

3.6 Daubechies Wavelets

For any $p \geq 1$, the Daubechies wavelet D_p has the shortest real orthogonal scaling function with approximation order p. It is not unique except for $p = 1$, but for any p there are only a finite number of solutions.

3.6.1 Bezout Approach

We first give a brief summary of the original derivation of Daubechies in [49] and [50], which is based on the Bezout equation.

We want to find all possible real orthogonal wavelets with p vanishing moments. This means that we need to find $h(\xi)$ of the form

$$h(\xi) = \left(\frac{1+e^{-i\xi}}{2}\right)^p h_0(\xi)$$

which satisfies

$$|h(\xi)|^2 + |h(\xi + \pi)|^2 = 1. \qquad (3.16)$$

Using

$$\frac{1+e^{-i\xi}}{2} = e^{-i\xi/2} \cos\frac{\xi}{2},$$

the orthogonality relation becomes

$$\left(\cos^2\frac{\xi}{2}\right)^p |h_0(\xi)|^2 + \left(\sin^2\frac{\xi}{2}\right)^p |h_0(\xi + \pi)|^2 = 1.$$

By assumption, h_0 has real coefficients. This makes $|h_0(\xi)|^2$ an even function, so we can write it in terms of cosines. Using the substitution

$$\cos\xi = 1 - 2\sin^2\frac{\xi}{2}$$

we find that

$$|h_0(\xi)|^2 = q(\sin^2\frac{\xi}{2}) \qquad (3.17)$$

for some real polynomial q. With the further substitution $y = \sin^2\xi/2$, equation (3.16) becomes

$$(1-y)^p q(y) + y^p q(1-y) = 1. \qquad (3.18)$$

This is a Bezout equation, whose general solution is given in equation (3.15). The fact that $q_1 = q_2 = q$ in this case adds the additional constraint

$$r(y) + r(1-y) = 0,$$

so r must have odd symmetry about $y = 1/2$.

The complete answer is

$$q(y) = \sum_{k=0}^{p-1} \binom{p-1+k}{k} y^k + y^p r(\frac{1}{2} - y), \qquad (3.19)$$

where r is an arbitrary odd polynomial.

Any q of this form will lead to a solution of equation (3.16) provided we can find an h_0 which satisfies equation (3.17). Solving (3.17) is called *spectral factorization*.

Chapter 3: Creating Wavelets

A necessary condition for the existence of a solution to equation (3.17) is obviously $q(y) \geq 0$ for $y \in [0, 1]$. A theorem of Riesz states that this is already sufficient. The solution itself can be found as follows.

Given $q(y)$, we reverse the substitution:

$$y = \sin^2 \frac{\xi}{2} = \frac{1}{2} - \frac{1}{2}\cos\xi = -\frac{1}{4}e^{-i\xi} + \frac{1}{2} - \frac{1}{4}e^{i\xi}, \qquad (3.20)$$

which turns $|h_0(\xi)|^2$ back into a trigonometric polynomial. With the standard substitution $z = e^{-i\xi}$, it turns into a Laurent polynomial with powers ranging from $-n$ to n, where n is the degree of q. We multiply $|h_0(z)|^2$ by z^n to turn it into a regular polynomial of degree $2n$.

It can then be shown that the roots of this polynomial come in groups of four, of the form z_k, $1/z_k$, z_k^*, $1/z_k^*$. If z_k is real or lies on the unit circle, there are only two elements in a group. We select one pair of the form z_k, $1/z_k^*$ from each group of four, or one element z_k from each group of two.

For simplicity, let z_k, $k = 1, \ldots, n$ be the chosen roots. Then

$$h_0(z) = \frac{(z - z_1) \cdots (z - z_n)}{(1 - z_1) \cdots (1 - z_n)} \qquad (3.21)$$

will be a solution of equation (3.17).

For a given number p of vanishing moments the shortest possible candidate for q is found by setting $r = 0$. These polynomials fortunately satisfy $q(y) \geq 0$ for $y \in [0, 1]$, so they lead to a family of orthogonal wavelets D_p, $p \geq 1$. D_p contains $2p$ coefficients; some books use the notation D_{2p} for that reason. These are the famous Daubechies wavelets.

Example 3.3

These are the calculations for D_2, probably the most-pictured wavelet of all. It was already shown in figure 1.6.

Formula (3.19) for $p = 2$ produces $q(y) = 1 + 2y$, which leads to

$$z|h_0(z)|^2 = -\frac{z^2}{2} + 2z - \frac{1}{2}.$$

The roots are $2 \pm \sqrt{3}$, a real pair of the form $z, 1/z$. If we take $z_1 = 2 + \sqrt{3}$, we get

$$h_0(z) = \frac{\sqrt{3} - 1}{2}(z - (2 + \sqrt{3})).$$

This leads to

$$h_0 = \frac{1 + \sqrt{3}}{4\sqrt{2}}, \quad h_1 = \frac{3 + \sqrt{3}}{4\sqrt{2}}, \quad h_2 = \frac{3 - \sqrt{3}}{4\sqrt{2}}, \quad h_3 = \frac{1 - \sqrt{3}}{4\sqrt{2}}.$$

Choosing z_1 to be the other root $2 - \sqrt{3}$ leads to the same coefficients in reverse order. □

One aspect of this derivation sometimes not mentioned is that the D_p are not unique except for $p = 1$ (which is the Haar wavelet). Different choices of roots in the spectral factorization lead to different wavelets.

Consistently choosing $1/z_k$ instead of z_k leads to the same recursion coefficients in reverse order. The scaling function is likewise reversed. For $p = 2$ (one pair of roots) and $p = 3$ (one quadruple of roots) this is the only non-uniqueness. For $p \geq 4$ there are several distinct solutions.

Daubechies herself produced three sets of Daubechies wavelets: her original choice (choose z_k inside the unit circle, in the upper half plane), the least asymmetric scaling functions, and the smoothest scaling functions. Other choices are possible.

The least asymmetric and smoothest functions are found by examining all possible solutions. The amount of work for doing that increases exponentially with p, but even for $p = 20$ it only amounts to a few hundred cases. Computers have no problem with that.

Incidentally, the coefficients for D_2 given in [49] and in the example above are produced by z_1 outside the unit circle, contrary to what the paper says elsewhere.

For $p = 1, 2, 3$ there are closed form expressions for the coefficients (see appendix A.) Exact expressions for the coefficients of D_4, D_5 in terms of radicals are given in [128]. The rest are only known numerically.

3.6.2 Projection Factor Approach

An alternative approach for deriving the Daubechies wavelets is based on the decomposition of the polyphase matrix in terms of projection factors. We use the factorization in equation (3.9) with Q chosen as in equation (3.11) and the projection factors parameterized as in equation (3.10). This provides approximation order 1 automatically. We can then impose additional approximation order conditions via constraints on the ν_k.

Example 3.4
We take $n = 1$, so that
$$P(z) = \frac{1}{\sqrt{2}} \begin{pmatrix} 1 & 1 \\ 1 & -1 \end{pmatrix} \frac{1}{\nu^2 + 1} \begin{pmatrix} \nu^2 + z & \nu(z-1) \\ \nu(z-1) & 1 + \nu^2 z \end{pmatrix}.$$

This will produce a scaling function with four recursion coefficients. We want two vanishing wavelet function moments.

The zeroth wavelet function moment is automatically 0. The first moment is
$$n_1 = \frac{1 - 3\nu^2}{2(\nu^2 + 1)},$$
so we must have
$$\nu = \pm \frac{1}{\sqrt{3}}.$$

Chapter 3: Creating Wavelets 79

The choice $\nu = -1/\sqrt{3}$ leads to the standard coefficients for D_2. The choice $\nu = 1/\sqrt{3}$ leads to the coefficients in reverse order. □

3.7 Coiflets

Coiflets are orthogonal wavelets for which ψ has several vanishing moments, and for which ϕ also has several vanishing moments (after the zeroth one).

Coiflets first appeared in [51]. They are named after Ronald Coifman, who requested such wavelets from Ingrid Daubechies. In Daubechies' original approach, the number of vanishing moments for ϕ and ψ was taken to be equal. Thus, we want

$$\int x^k \psi(x)\, dx = 0, \qquad k = 0, \ldots, p-1,$$

$$\int \phi(x)\, dx = 1,$$

$$\int x^k \phi(x)\, dx = 0, \qquad k = 1, \ldots, p.$$

The advantage of coiflets is that for smooth signals $s(x)$, the scaling function expansion coefficients s_{nk} are very close to $s(2^{-n}k)$. The discussion in section 2.2 showed that in general

$$|s_{nk} - 2^{-n/2} s(2^{-n}k)| = O(2^{-n}),$$
$$|s_{nk} - 2^{-n/2} s(2^{-m}(k + \mu_1))| = O(2^{-2n}),$$

but for coiflets we have

$$|s_{nk} - 2^{-n/2} s(2^{-n}k)| = O(2^{-(p+1)n}).$$

3.7.1 Bezout Approach

We first look at Daubechies' original approach to constructing coiflets.

For a given p, we want ψ to have p vanishing moments, which means

$$h(\xi) = \left(\frac{1 + e^{-i\xi}}{2}\right)^p h_0(\xi) = e^{-ip\xi/2}\left(\cos\frac{\xi}{2}\right)^p h_0(\xi),$$

and we also want $\mu_1, \ldots, \mu_p$ to vanish, which means

$$h(\xi) = 1 + \left(\frac{1 - e^{-i\xi}}{2}\right)^p h_1(\xi) = 1 + \left(ie^{-i\xi/2}\right)^p \left(\sin\frac{\xi}{2}\right)^p h_1(\xi).$$

For even $p = 2n$, we assume that we can write
$$e^{-in\xi} h_0(\xi) = q_1(y),$$
$$e^{-in\xi} h_1(\xi) = q_2(y),$$
where $y = \sin^2(\xi/2)$ as in the derivation of the Daubechies wavelets. This leads to the Bezout equation
$$(1-y)^n q_1(z) + (-1)^n y^n q_2(1-y) = 1,$$
whose solution we know:
$$q_1(y) = \sum_{k=0}^{n-1} \binom{n+k-1}{k} y^k + y^n r(y). \tag{3.22}$$

A similar construction can be used for odd p.

There are some differences between the Daubechies wavelet construction and the coiflet construction:

- In the coiflet construction, q_1 and q_2 are unrelated, so there is no restriction on r. In fact, we take r to be arbitrary *after* we turn q_1 back into a trigonometric polynomial. It does not have to be of the form $r(\sin^2(\xi/2))$. This means that in equation (3.22), q_1 is not really a polynomial.

- In the coiflet construction, the solution of the Bezout equation gives us h_0 directly. There is no need for spectral factorization.

- The Daubechies wavelet construction had the orthogonality relation built in. The coiflet construction does not. This means we have to enforce orthogonality by imposing conditions on r.

Example 3.5
We consider the easiest case $p = 2$. This leads to
$$q_1(z) = 1 + zr(z),$$
which means
$$e^{-i\xi} h_0(\xi) = 1 + \frac{1 - \cos \xi}{2} r(\xi).$$
We assume
$$r(\xi) = a + be^{-i\xi}$$
and calculate
$$1 = |h(\xi)|^2 + |h(\xi + \pi)|^2$$
$$= \frac{1}{256} \left[2(a^2 + b^2 - 4b) \cos 4\xi - 8(a^2 + b^2 + 4a - 4) \cos 2\xi \right.$$
$$\left. + 4(3a^2 + 3b^2 + 16a + 8b + 48) \right].$$

Chapter 3: Creating Wavelets

The resulting equations for a and b have six solutions. The two real solutions are

$$a = \frac{-1 \pm \sqrt{7}}{2}, \qquad b = 1 - a.$$

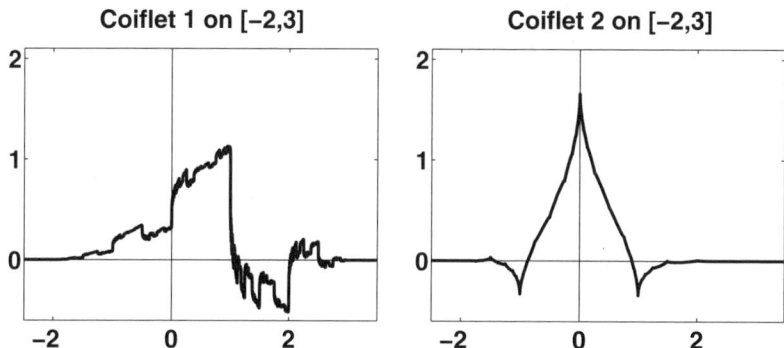

FIGURE 3.1
The two different coiflets on $[-2, 3]$.

The resulting coefficients are listed in appendix A. These functions have support in $[-2, 3]$. The two solutions are quite different (fig. 3.1).

If we use instead the assumption

$$r(\xi) = ae^{-i\xi} + be^{-2i\xi},$$

we again find two real solutions

$$a = \frac{-1 \pm \sqrt{15}}{2}, \qquad b = 1 - a.$$

These functions have support in $[-1, 4]$. Figure 3.2 shows one of them; the second one is a highly discontinuous L^2-function.

We can likewise find two coiflets with support on $[-3, 2]$ (which are the reverses of those on $[-2, 3]$) and on $[-4, 1]$ (which are the reverses of those on $[-1, 4]$). □

3.7.2 Projection Factor Approach

As in the case of the Daubechies wavelets, we can use the projection factor decomposition instead.

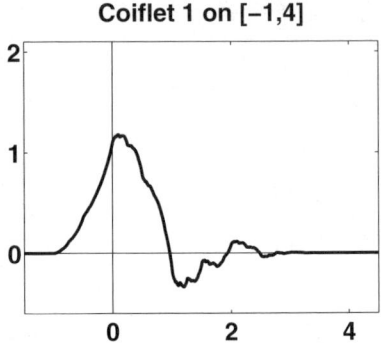

FIGURE 3.2
One of the two coiflets on $[-1, 4]$.

Example 3.6

Assume we are looking for the coiflets with two vanishing moments, of length 6. The general parametrization for the polyphase matrix of an orthogonal wavelet of length 6 with approximation order 1 is

$$P(z) = \frac{1}{\sqrt{2}} \begin{pmatrix} 1 & 1 \\ 1 & -1 \end{pmatrix} \frac{1}{(\nu_1^2 + 1)(\nu_2^2 + 1)} \\ \times \begin{pmatrix} \nu_1^2 + z & \nu_1(z-1) \\ \nu_1(z-1) & 1 + \nu_1^2 z \end{pmatrix} \begin{pmatrix} \nu_2^2 + z & \nu_2(z-1) \\ \nu_2(z-1) & 1 + \nu_2^2 z \end{pmatrix}. \qquad (3.23)$$

We convert this to the symbol, and add a factor of $e^{2i\xi}$ to shift the support to $[-2, 3]$.

We then compute the moments n_1 and m_1 and set them to zero. This gives us two quadratic equations in ν_1, ν_2. The two real solutions are

$$\nu_1 = \frac{1}{3}\left(-2 \pm \sqrt{7}\right), \qquad \nu_2 = -2 \mp \sqrt{7},$$

which lead back to the same answers as before.

If we add a factor of $e^{i\xi}$ instead, we get the solutions on $[-1, 4]$.

It is not necessary to add the condition $m_2 = 0$. Theorem 3.3 below shows that it comes for free. □

3.7.3 Generalized Coiflets

More generally, we could look for coiflets where ϕ and ψ have a different number of vanishing moments. The projection factor approach can handle this easily. The Bezout approach could probably be adapted as well.

Given that we want p vanishing moments for ψ and q vanishing moments for ϕ, what is the minimum length possible for a coiflet?

Chapter 3: Creating Wavelets

At first glance, the answer appears to be $2(p+q)$: each increase of 2 in the length of a scaling function gives us two new parameters and one new orthogonality condition to be satisfied, so there is a net gain of one degree of freedom.

In fact, a shorter length suffices. For the standard coiflets with $p = q$, in particular, length $3p$ instead of $4p$ is sufficient.

The reason for that is the following theorem.

THEOREM 3.3
Assume that ϕ, ψ form a real orthogonal wavelet, and ψ has p vanishing moments. If
$$\mu_{2k-1} = 0, \quad k = 0, \ldots, p-1,$$
then also
$$\mu_{2k} = 0, \quad k = 0, \ldots, p-1.$$

PROOF By lemma 1.27 it is sufficient to prove this theorem for the discrete moments m_k instead of the continuous moments μ_k.

If ϕ has p vanishing moments, then h has a zero of order p at π; so $|h|^2$ has a zero of order $2p$ at π.

By repeatedly differentiating the orthogonality relation
$$|h(\xi)|^2 + |h(\xi + \pi)|^2 = 1,$$
we see that
$$(D^k |h|^2)(0) = 0, \quad k = 1, \ldots, 2p-1.$$
Choose any $k \leq p-1$ and assume we already know that $\mu_1 = \cdots = \mu_{2k-1} = 0$. Then

$$0 = \left(D^{2k}|h|^2\right)(0) = \sum_{\ell=0}^{2k} \binom{2k}{\ell} D^\ell h(0) \left(D^{2k-\ell} h(0)\right)^*$$

$$= (-1)^k \sum_{\ell=0}^{2k} \binom{2k}{\ell} (-1)^\ell m_\ell m^*_{2k-\ell}$$

$$= (-1)^k 2 m_{2k}. \qquad \blacksquare$$

An alternative proof is given in [31].

This theorem says that if the wavelet function has p vanishing moments, then we only need to prescribe the scaling function moments of *odd* order to be zero (up to a certain order); the moments of even order will automatically be zero.

The following table gives an (incomplete) list of the possible patterns of zeros for coiflets of a certain length L:

	L = 4		6		8		10		12	
	μ_k	ν_k	μ_k	ν_k	μ_k	ν_k	μ_k	ν_k	μ_k	ν_k
k = 0	1	0	1	0	1	0	1	0	1	0
1	0	×	0	0	0	0	0	0	0	0
2	×	×	<u>0</u>	×	<u>0</u>	0	<u>0</u>	0	<u>0</u>	0
3			×	×	×	×	0	×	0	0
4							<u>0</u>	×	<u>0</u>	×
5							×	×	×	×

× indicates a nonzero number. The underlined zeros come for free.

An excellent discussion of the available choices, with many literature references, can be found in section 6.9 of [30], which is largely the same as the paper [31]. The authors distinguish three standard series of coiflets with approximately equal numbers of vanishing moments for ϕ and ψ:

length	p	q prescribed	q actual	comparison
$6n$	$2n$	n	$2n$	$q = p$
$6n + 2$	$2n + 1$	n	$2n$	$q = p - 1$
$6n + 4$	$2n + 1$	$n + 1$	$2n + 2$	$q = p + 1$

Instead of demanding vanishing moments of ϕ at $x = 0$, we could choose a different point τ and demand that

$$\mu_0 = 1,$$
$$\mu_k = \tau^k, \quad k = 1, \ldots, p.$$

This has the effect that

$$s_{nk} \approx s(2^{-m}(k + \tau))$$

to high accuracy.

This is just a shift. If we choose support $[0, 5]$ and $\tau = 2$, we get the two coiflets on $[-2, 3]$, shifted to the right by 2. However, it also opens the possibility of using noninteger τ. This approach is taken in [118].

Example 3.7
We look at the two-parameter family in equation (3.23), and impose the conditions $n_2 = 0$ and $m_1 = \tau$ (which automatically implies $m_2 = \tau^2$).

A pair of real solutions exists for any τ in the range

$$\frac{5 - \sqrt{15}}{2} \leq \tau \leq \frac{5 + \sqrt{15}}{2}.$$

This includes the integers 1 through 4, which correspond to the coiflets with support $[-1, 4]$ through $[-4, 1]$ derived before.

We can use a root finder to see if there are any τ for which we get $\mu_3 = \tau^3$. There are two of them: $\tau_1 \approx 2.1059678$, and $\tau_2 = 5 - \tau_1$. Each of them leads to two distinct coiflets. The coiflets for τ_1 look pretty similar to the coiflets with $\tau = 2$. The coiflets for τ_2 are the same functions in reverse. □

Chapter 3: Creating Wavelets   85

3.8 Cohen Wavelets

Finding biorthogonal wavelets is much easier than finding orthogonal wavelets. As outlined in the discussion of the completion problem above, we can start with almost any refinable function and find a dual by solving a linear system of equations.

We will just discuss one particular family of biorthogonal wavelets, the Cohen family derived in [40]. The Cohen$(p, \tilde{p})$ wavelet is characterized by the following properties:

- The scaling function ϕ is the B-spline of order p. This is a refinable function with symbol

$$h(\xi) = \left(\frac{1 + e^{-i\xi}}{2}\right)^p.$$

- The dual scaling function $\tilde{\phi}$ has approximation order $\tilde{p}$, is symmetric about the same point as ϕ, and is as short as possible.

These wavelets exist for each $(p, \tilde{p})$ with $p, \tilde{p} \geq 1$, p and $\tilde{p}$ both even or both odd.

Cohen wavelets were derived in [40], using a Bezout-type approach, but they can be constructed more simply by shifting factors in the Daubechies wavelets around.

To construct the Cohen$(p, \tilde{p})$ wavelet, we begin with the Daubechies wavelet D_n, $n = (p + \tilde{p})/2$:

$$h(\xi) = \left(\frac{1 + e^{-i\xi}}{2}\right)^n h_0(\xi).$$

We shift some of the approximation order factors and all of the h_0 over to the other side, and add an exponential term on both sides to center them at 0 (for even p) or $1/2$ (for odd p). For even p, we get

$$h_{\text{new}}(\xi) = e^{ip\xi/2} \left(\frac{1 + e^{-i\xi}}{2}\right)^p,$$

$$\tilde{h}_{\text{new}}(\xi) = e^{i\tilde{p}\xi/2} \left(\frac{1 + e^{-i\xi}}{2}\right)^{\tilde{p}} |h_0(\xi)|^2.$$

For odd p, the factor in front is $e^{i(p-1)\xi/2}$ instead.

The resulting scaling functions have all the required properties.

Example 3.8
We start with the scaling function of the Daubechies wavelet D_3. We do not actually need to know the symbol; we just need to know $|h(\xi)|^2$, which is

much easier:
$$|h(\xi)|^2 = \left(\cos\frac{\xi}{2}\right)^6 |h_0(\xi)|^2$$
with
$$|h_0(\xi)|^2 = \frac{1}{8}\left[3e^{-2i\xi} - 18e^{-i\xi} + 38 - 18e^{i\xi} + 3e^{2i\xi}\right].$$

This is the polynomial $q(y)$ for $p = 3$ from equation (3.19), after applying the reverse substitution in equation (3.20).

To find the Cohen(1,5) wavelet, we divide this as
$$h(\xi) = \frac{1 + e^{-i\xi}}{2} \qquad \text{(Haar scaling function)},$$
$$\tilde{h}(\xi) = e^{2i\xi}\left(\frac{1 + e^{-i\xi}}{2}\right)^5 |h_0(\xi)|^2.$$

For the Cohen(2,4) wavelet, we divide it as
$$h(\xi) = e^{i\xi}\left(\frac{1 + e^{-i\xi}}{2}\right)^2 \qquad \text{(hat function)},$$
$$\tilde{h}(\xi) = e^{2i\xi}\left(\frac{1 + e^{-i\xi}}{2}\right)^4 |h_0(\xi)|^2.$$

The coefficients of this wavelet and graphs of all four functions are given in example 1.6.

For the Cohen(3,3) wavelet, we divide the factors as
$$h(\xi) = e^{i\xi}\left(\frac{1 + e^{-i\xi}}{3}\right)^3 \qquad \text{(quadratic B-spline)},$$
$$\tilde{h}(\xi) = e^{i\xi}\left(\frac{1 + e^{-i\xi}}{2}\right)^3 |h_0(\xi)|^2. \qquad \square$$

3.9 Other Constructions

Many other kinds of wavelets are possible. Daubechies' book mentions a number of them. Other examples are given in [18], [44], [60], [62], [66], [78], and [119].

We will just mention one more possibility. We go through the construction of Daubechies wavelets until we get to the part where we sort the roots of the polynomial into real pairs and complex quadruples. Then we assign some of the roots to h_0, and some to $\tilde{h}_0$, and keep the approximation orders even.

Chapter 3: Creating Wavelets

To ensure that the result is real, we cannot do this arbitrarily: we can shift real roots any way we want, but complex roots have to be shifted in pairs of the form z_k, $1/z_k^*$.

For $p = 2$ we get nothing new. There are two real roots. Either we split them up, and get the standard Daubechies wavelet D_2; or we put them both on one side, and get Cohen(2,2).

For $p = 3$, we also get nothing new. There is a complex quadruple, and we either get D_3 or Cohen(3,3).

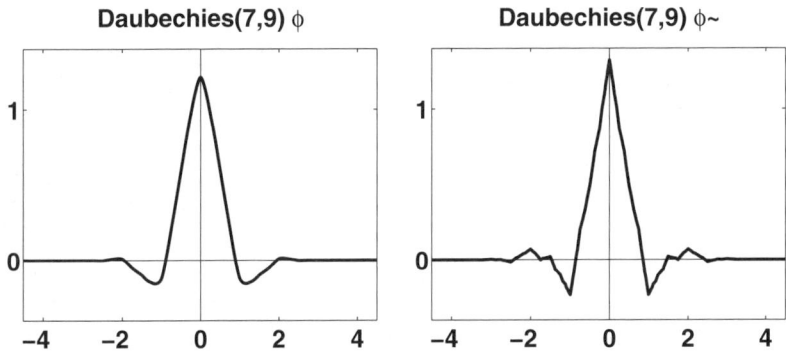

FIGURE 3.3
Scaling functions of biorthogonal Daubechies(7,9) pair

The first new kind appears for $p = 4$. The roots consist of a real pair and a complex quadruple. To achieve the closest similarity in length, we put the two real roots on one side, and the four complex roots on the other. This produces the Daubechies(7,9) wavelet, named for the number of coefficients (fig. 3.3).

4
Applications

Wavelet-based algorithms have found many uses in different areas. Most of them are related to signal processing or numerical analysis. We give a brief description of some of the main applications in this chapter.

4.1 Signal Processing

The fundamental operation in signal processing is called *filtering* by the engineers and *convolution* by the mathematicians. A signal $s(x)$ is converted to

$$(f * s)(x) = \int f(y-x)s(x)\,dx,$$

or in the discrete setting

$$(f * s)_n = \sum_k f_{n-k} s_k.$$

Since

$$\widehat{(f * s)}(\xi) = \sqrt{2\pi}\,\hat{f}(\xi)\hat{s}(\xi),$$

this corresponds to selecting a segment of the Fourier transform $\hat{s}(\xi)$ determined by the cutoff function $\hat{f}$. This segment is called a *subband*.

If $\hat{f}$ is localized near 0, it is called a *low-pass* filter. If $\hat{f}$ is centered elsewhere, it is a *high-pass* or *band-pass* filter.

A basic signal processing algorithm consists of three parts:

- Decompose the original signal into subbands.

- Do some processing on the subbands.

- Recombine the subbands into the processed signal.

The discrete wavelet transform (DWT) represents one particular kind of decomposition and reconstruction.

4.1.1 Detection of Frequencies and Discontinuities

The DWT decomposes a signal into its frequency components, just like the fast Fourier transform (FFT), but it localizes the frequency components in time. We can easily detect *when* a particular frequency band is present.

Many features of the signal can be detected in the DWT. For example, a singularity in the signal leads to large d-coefficients at all levels, sharply localized in time. White noise produces large d-coefficients at many levels over longer periods. Smooth frequency content shows up at one or two levels corresponding to the frequency.

This can be done for one-dimensional or two-dimensional signals. We just show one-dimensional examples in this section.

Example 4.1
The left part of figure 4.1 shows the wavelet decomposition of a signal over three levels. The numbers are the same as in figure 2.1, but the d-coefficients have been magnified to show better.

The signal has a jump near the beginning, followed by a smooth wave of increasing frequency. The singularity and the frequencies in the smooth part are visible in the DWT, as described above.

The same signal with added noise is shown in the left part of figure 4.2. The noise is visible at all levels. The d-coefficients in this picture have not been magnified. It should be compared to figure 2.1 rather than figure 4.1.
☐

4.1.2 Signal Compression

The DWT concentrates much of the energy of a signal in a few large coefficients. A generally smooth signal with localized nonsmooth features is transformed into relatively few large s-coefficients which describe the overall shape, a few larger d-coefficients which describe the local features, and a lot of very small d-coefficients.

If we set small coefficients below some threshold to zero, the reconstructed signal will be quite close in shape to the original.

For added compression, the larger coefficients could also be quantized. Quantization basically means that we round each number to one with fewer decimals or bits.

Wavelet-based compression algorithms have been incorporated into the FBI WSQ algorithm for fingerprint image compression [28] and into the JPEG-2000 standard.

Example 4.2
Figure 4.1 shows the decomposition of a signal using the Daubechies(7,9) wavelet over three levels on the left. The right side shows the d-coefficients

Chapter 4: Applications

after thresholding, and the reconstructed approximate signal.

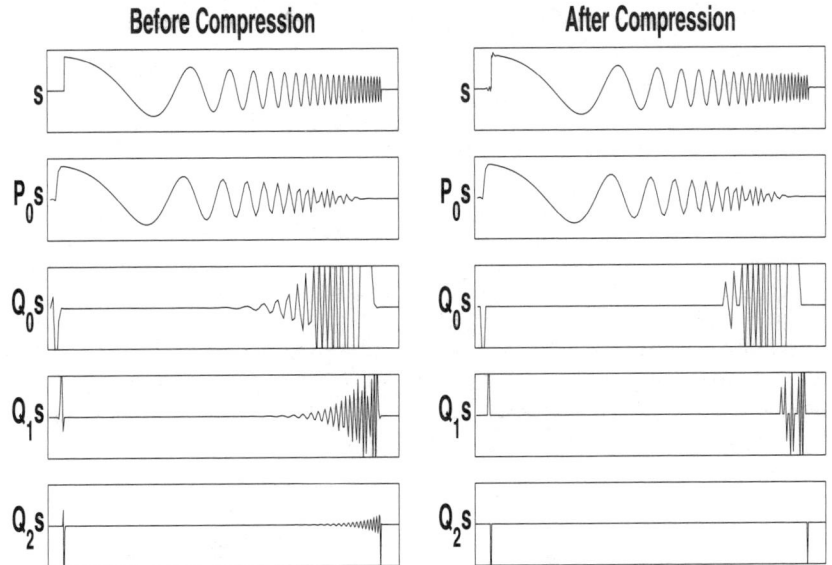

FIGURE 4.1
Left: Decomposition of a signal over three levels. Right: Subbands after thresholding, and reconstructed signal.

The d-coefficients are magnified to make the compression effect more visible. The threshold was selected to result in a compression ratio of about 7:1. □

4.1.3 Denoising

Random noise in a signal will show up mostly in the d-coefficients. If we set the smaller coefficients to zero, much of the noise will disappear (along with minor features of the signal, but that cannot be avoided). The process is the same as for compression, but with a different purpose.

Wavelet-based denoising was proposed and analyzed in great detail by Donoho [55] to [58], and by others. It is often referred to as *wavelet shrinkage*.

A *threshold function* T_ϵ is applied to each d-coefficient:

$$d_{nk} \to T_\epsilon(d_{nk}).$$

The most commonly used methods are *hard thresholding*

$$T_\epsilon(x) = \begin{cases} 0 & \text{if } |x| \leq \epsilon \\ x & \text{otherwise} \end{cases}$$

and *soft thresholding*

$$T_\epsilon(x) = \begin{cases} x - \epsilon & \text{if } x > \epsilon \\ 0 & \text{if } |x| \leq \epsilon \\ -x + \epsilon & \text{if } x < -\epsilon \end{cases}.$$

Example 4.3
White noise with a standard deviation of about 15% of signal amplitude has been added to the signal. Figure 4.2 shows the wavelet decomposition of the noisy signal over three levels on the left, and the reconstructed signal and subbands after hard thresholding on the right.

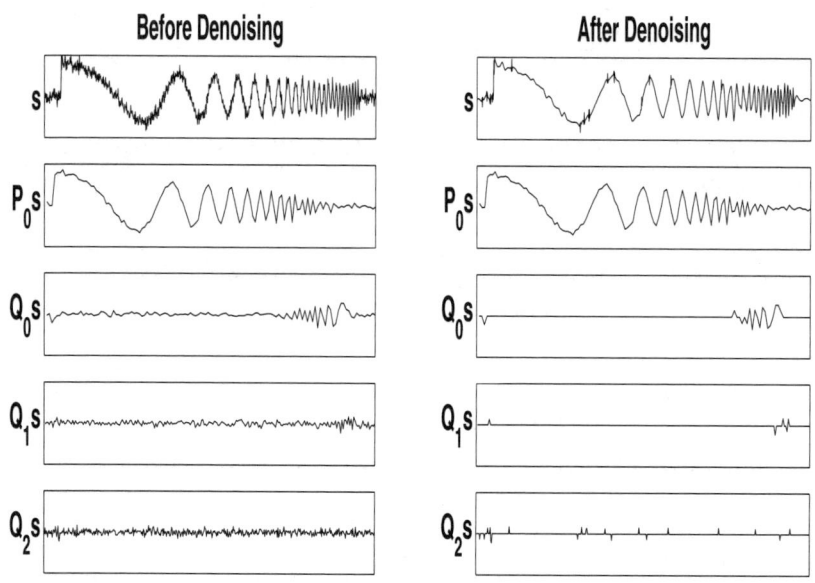

FIGURE 4.2
Left: Decomposition of a noisy signal over three levels. Right: Subbands after thresholding, and reconstructed signal.

□

Chapter 4: Applications

It is observed in the literature that denoising by thresholding works better if several shifted copies of the signal are denoised and averaged. The denoising algorithm is not shift invariant because the DWT is not shift invariant.

Consider one DWT step:

original signal	s_{n0}	s_{n1}	s_{n2}	s_{n3}	...
convolved signal	$(h*s_n)_0$	$(h*s_n)_1$	$(h*s_n)_2$	$(h*s_n)_3$	...
DWT of signal	$s_{n-1,0}$		$s_{n-1,1}$		
DWT of shifted signal		$s_{n-1,0}$		$s_{n-1,1}$	

If we shift the signal by one, we get different coefficients at the next lower level. If we shift the signal by two, we get the original coefficients again, shifted by one. After k levels, we need a shift by 2^k before the original coefficients at the lowest level reappear.

Denoising several shifted copies of the signal and averaging them improves the results. This can be implemented efficiently by doing a decomposition *without* downsampling, thresholding the coefficients, and doing an averaged reconstruction. The amount of work needed for a signal of length N over k levels is $O(kN)$, instead of $O(2^k N)$ when we denoise all possible shifts separately.

4.2 Numerical Analysis

4.2.1 Fast Matrix–Vector Multiplication

Assume we have a matrix T whose entries t_{jk} vary smoothly with j,k, except maybe for some locations. Matrices of this type come up, for example, in evaluating integral operators. The discontinuities are often near the diagonal.

When we do a DWT on the rows and columns of T, the resulting matrix will have large values in one corner, mostly small values in the rest of the matrix. This is really the same as image processing. We then set small values to zero and end up with a sparse matrix. To take advantage of that, we have to decompose the vector as well.

For simplicity, assume we are using orthogonal wavelets. Decomposing T is equivalent to multiplying T with the transform matrix L on the left and L^* on the right. To evaluate
$$T\mathbf{x} = \mathbf{y},$$
we do instead
$$(LTL^*)(L\mathbf{x}) = L\mathbf{y}.$$

In words, we transform the matrix, transform the vector $\mathbf{x}$, multiply the two, and do a reconstruction on the result. The final result is not quite correct, of course, but it will be close.

A small threshold produces a small error, but the matrix is not very sparse. For a larger threshold, the matrix is sparser, but the error grows. This is mostly useful for very large problems where accuracy is not as important as speed.

Example 4.4

We take a 128 × 128 matrix with entries

$$t_{jk} = \frac{1}{(|j-k|+1)^2}.$$

The entries are large on the diagonal, and rapidly fall off in a smooth manner.

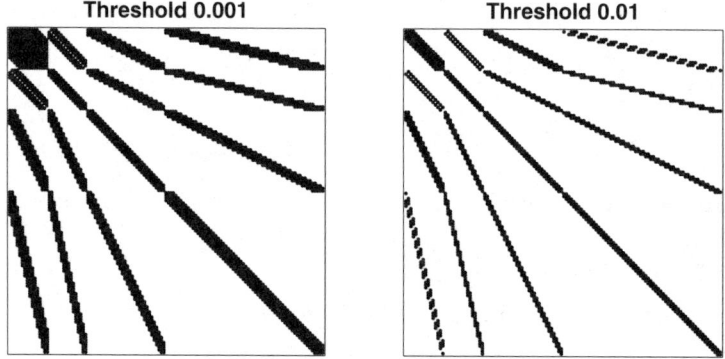

FIGURE 4.3
Sparsity pattern of a transformed matrix after thresholding.

After three levels of decomposition with the Haar wavelet, we set small coefficients to zero. The resulting matrix has a distinct sparsity pattern: bands radiating out from one corner (fig. 4.3.) □

This process has been analyzed in detail for certain kinds of matrices in [24]. Other papers on this subject are [7], [77], and [97].

The amount of work for an $N \times N$ fast matrix–vector multiplication is $O(N \log N)$, compared to the usual $O(N^2)$. Using another trick called *nonstandard decomposition* the work can be reduced to $O(N)$. (Nonstandard decomposition involves saving and using the s-coefficients at all levels, not just the lowest level.)

Chapter 4: Applications 95

4.2.2 Fast Operator Evaluation

Fast operator evaluation involves representing operators directly in terms of the wavelet coefficients.

Assume T is any kind of linear operator. For simplicity, we assume that the wavelet is orthogonal, and that we begin at level 0. We are given a function f, and want to compute Tf.

The function f is expanded in a scaling function series

$$f \approx \sum_k f_k^* \phi_k,$$

where $\phi_k(x) = \phi(x - k)$.

We precompute

$$T\phi_k \approx \sum_j a_{kj}^* \phi_j.$$

The a_{kj} are called the *connection coefficients*. If T is translation invariant, then $a_{kj} = a_{k-j}$. In general,

$$a_{kj}^* = \langle T\phi_k, \phi_j \rangle.$$

Then

$$Tf \approx T\left(\sum_k f_k^* \phi_k\right)$$

$$\approx \sum_{kj} f_k^* a_{kj}^* \phi_j$$

$$= \sum_k \left(\sum_j f_k^* a_{kj}^*\right) \phi_j,$$

so the scaling function expansion coefficients of Tf are given approximately by

$$(Tf)_k^* \approx \sum_j f_k^* a_{kj}^*.$$

There are two reasons why this is useful. First, if we have an equation involving differential or integral operators, we can expand the coefficient functions and the unknown solution in terms of a scaling function series, and discretize the equation completely in terms of the coefficients. This is a Galerkin method based on the shifted scaling functions as basis functions.

Second, we now have the multiresolution properties of wavelets at our disposal. The effect of this will be discussed in the next section.

Details of this approach have been worked out for derivative operators, the Hilbert transform, shifts, and multiplication and powers of functions [21], [22],

[85], [86], and [112]. Jameson [84] also worked out the connection coefficients for differentiation for the boundary wavelets of [42].

Example 4.5
If T is the derivative operator, the connection coefficients are
$$a_{kj} = \langle \phi'(x-k), \phi(x-j) \rangle = a_{k-j},$$
which can be evaluated with the methods of section 2.8.1.

For the scaling function of the Daubechies wavelet D_2, the result is
$$(a_{-2}, a_{-1}, a_0, a_1, a_2) = (-\frac{1}{12}, \frac{2}{3}, 0, -\frac{2}{3}, \frac{1}{12}). \qquad \square$$

4.2.3 Differential and Integral Equations

Wavelets can be used as basis functions for Galerkin methods, as already mentioned in the previous section. If we just do this at one level, they have no particular advantage over other types of functions. Their real power lies in the multiresolution approach.

The effect is quite different for differential and for integral equations. Integral equations typically lead to full matrices even if the basis functions are localized. However, these matrices tend to have smoothly varying entries where the kernel of the integral operator is smooth, so we can use the techniques for fast matrix–vector multiplication in section 4.2.1 above.

For differential equations, finite element matrices are already sparse, but they tend to be ill-conditioned. Wavelet decomposition increases the bandwidth of the matrices a bit, but they are still sparse. It has been found that after decomposition the coefficients at each level can be scaled by an appropriate constant to improve the conditioning of the matrix. This is called *multilevel* or *hierarchical preconditioning*. It will speed up iterative algorithms such as the conjugate gradient method.

It is hard work, however, to compute the coefficients (often by evaluating integrals as described in section 2.8) and to handle boundaries, especially in higher dimensions. For this reason, wavelet methods in partial differential equations (PDEs) have not become as popular as initially predicted.

It appears that what is important is the multilevel structure, not the wavelets and the DWT *per se*. There are other methods to get from level to level which are easier to implement.

There is an extensive list of articles dealing with wavelet methods in numerical analysis. We content ourselves with mentioning some survey papers and books.

For wavelets in numerical analysis in general, see [23] and [25], and the book chapter [37], later expanded into book [38].

For PDEs, see [15], [26], [32], [33], and [83], and the books [47] and [101].

For integral equations, see [35], [102], and [116].

5

Existence and Regularity

Previous chapters explained the basic ideas behind refinable functions, multiresolution approximations (MRAs), and the discrete wavelet transform (DWT), as well as ways for determining some basic properties of the basis functions: approximation order, moments, and point values.

These results were all presented under the assumption that the underlying refinement equation defines a scaling function ϕ with some minimal regularity properties, and that this function produces an MRA with a wavelet function ψ. In the biorthogonal case, we also assumed the existence of $\tilde{\phi}$ and $\tilde{\psi}$.

In order for the DWT algorithm to work, it is actually not necessary that such a ϕ really exists: if we have sets of recursion coefficients which satisfy the biorthogonality conditions in equation (1.11), they will give rise to a DWT algorithm that works on a purely algebraic level. We may not be able to justify the interpretation of the DWT as a splitting of the original signal into a coarser approximation and fine detail at different levels; there may also be numerical stability problems when we decompose and reconstruct over many levels, but the algorithm will be an invertible transform.

In this chapter, we will give necessary and sufficient conditions for existence, regularity, and stability of ϕ and ψ. The material is rather mathematical in nature.

We note that to establish existence and regularity, it suffices to look at the scaling function ϕ. The wavelet function ψ is just a finite linear combination of scaled translates of ϕ, so it automatically inherits those properties. We only need to look at ψ to check its stability.

There are two main approaches: in the time domain (section 5.4) and in the frequency domain (sections 5.1 to 5.3.) The time domain approach is based on the refinement equation

$$\phi(x) = \sqrt{2} \sum_{k=k_0}^{k_1} h_k \, \phi(2x - k); \tag{5.1}$$

the frequency domain approach is based on the Fourier transform of equation (5.1), which is

$$\hat{\phi}(\xi) = h(\xi/2)\hat{\phi}(\xi/2), \tag{5.2}$$

97

where $h(\xi)$ is the symbol of ϕ, defined by

$$h(\xi) = \frac{1}{\sqrt{2}} \sum_{k=k_0}^{k_1} h_k e^{-ik\xi}.$$

5.1 Distribution Theory

Ultimately we are interested in solutions of the refinement equation (5.1), but at first we will consider solutions of equation (5.2). We want $\hat{\phi}$ to be a function, but ϕ itself could be a distribution.

THEOREM 5.1
A necessary condition for the existence of a function $\hat{\phi}$ which is continuous at 0 with $\hat{\phi}(0) \neq 0$ and which satisfies equation (5.2) is that $h(0) = 1$.
 This condition is also sufficient: if $h(0) = 1$, then

$$\Pi_n(\xi) = \prod_{k=1}^{n} h(2^{-k}\xi) \tag{5.3}$$

converges uniformly on compact sets to a continuous limit function Π_∞ with polynomial growth.
 For any nonzero choice of $\hat{\phi}(0)$,

$$\hat{\phi}(\xi) = \Pi_\infty(\xi)\hat{\phi}(0)$$

is the unique continuous solution of equation (5.2) with given value at 0, and it is the Fourier transform of a distribution ϕ with compact support in the interval $[k_0, k_1]$.

PROOF The necessity is easy. For $\xi = 0$, equation (5.2) says

$$\hat{\phi}(0) = h(0)\hat{\phi}(0).$$

If $\hat{\phi}(0) \neq 0$, this is only possible if $h(0) = 1$.
 Assume now that $h(0) = 1$. $h(\xi)$ is a trigonometric polynomial, so it is bounded and differentiable. We choose constants c, d so that

$$|h(\xi)| \leq 1 + c|\xi| \leq e^{c|\xi|},$$
$$|h(\xi)| \leq 2^d.$$

Then for all n,

$$|\Pi_n(\xi)| \leq e^{c\sum_{k=1}^{n}|2^{-k}\xi|} < e^{c|\xi|}.$$

Chapter 5: Existence and Regularity 99

In particular, $|\Pi_n(\xi)|$ is uniformly bounded by e^c for $|\xi| \leq 1$.

For arbitrary ξ, choose N so that $2^{N-1} \leq |\xi| < 2^N$. For $n \geq N$,

$$|\Pi_n(\xi)| = \left|\prod_{k=1}^{N-1} h(2^{-k}\xi)\right| \cdot \left|\prod_{k=N}^{n} h(2^{-k}\xi)\right| \qquad (5.4)$$
$$\leq 2^{d(N-1)} e^c$$
$$\leq |\xi|^d e^c.$$

This establishes polynomial growth.

For $m > n \geq N$,

$$|\Pi_m(\xi) - \Pi_n(\xi)| = |\Pi_n(\xi)| \cdot \left|1 - \prod_{k=n+1}^{m} h(2^{-k}\xi)\right|$$
$$\leq |\xi|^d e^c \left|1 - e^{c \cdot 2^{-n}|\xi|}\right|$$
$$\leq |\xi|^d e^c \, c \, 2^{-n}|\xi|.$$

This goes to zero uniformly for all ξ in any compact interval, so $\Pi_n(\xi)$ converges uniformly on compact sets as $n \to \infty$. Since each Π_n is continuous, the limit function is continuous.

If $\hat{\phi}$ is any solution of equation (5.2) which is continuous at 0, then

$$\hat{\phi}(\xi) = \Pi_n(\xi)\hat{\phi}(2^{-n}\xi)$$
$$= \Pi_n(\xi)\left[\hat{\phi}(2^{-n}\xi) - \hat{\phi}(0)\right] + \Pi_n(\xi)\hat{\phi}(0).$$

As $n \to \infty$ for ξ in any compact interval, the first term on the right goes to zero, and the second term converges to the continuous function with polynomial growth

$$\hat{\phi}(\xi) = \Pi_\infty(\xi)\hat{\phi}(0).$$

This establishes uniqueness.

To show compact support, we use a technique from [79].

$$\hat{\phi}_n(\xi) = \Pi_n(\xi)\hat{\phi}(0)$$

is a linear combination of terms of the form $e^{-i 2^{-n} k \xi}$ with $k_0 \leq 2^{-n} k \leq k_1$.

Its inverse Fourier transform ϕ_n is a linear combination of δ-distributions at the points $2^{-n} k$. Since $\hat{\phi}_n$ converges to $\hat{\phi}$ uniformly on compact sets, ϕ_n converges weakly to ϕ. Each ϕ_n has support in $[k_0, k_1]$, so ϕ must have its support in the same interval. ∎

REMARK 5.2 The preceding proof shows that the given solution ϕ is the only one whose Fourier transform is a function continuous at 0 with $\hat{\phi}(0) \neq 0$. The uniqueness disappears if we drop any of these assumptions.

There may be other distribution solutions with $\hat{\phi}(0) = 0$, there may be solutions with discontinuous $\hat{\phi}$, and there may even be distribution solutions which have no Fourier transform. For example, the Hilbert transform of any tempered distribution solution is also a solution. ∎

Theorem 5.1 can be generalized by removing the requirement that $\hat{\phi}(0) \neq 0$.

THEOREM 5.3
A necessary condition for the existence of a compactly supported distribution solution of the refinement equation is that $h(0) = 2^n$ for some integer $n \geq 0$.

If $n > 0$, then $\hat{\phi}(0) = 0$, and ϕ is the nth derivative of a distribution solution Φ of

$$\Phi(x) = 2^{-n}\sqrt{2}\sum_{k=a}^{b} h_k \Phi(2x - k).$$

A proof can be found in [79].

5.2 L^1-Theory

In this section we will look for sufficient conditions for the existence of an L^1-solution of the refinement equation. As a by-product, we will also get smoothness estimates for the solution.

We already know that the condition $h(0) = 1$ is enough to guarantee the existence of $\hat{\phi}$. To show that this function has an inverse Fourier transform which is a function, we need to impose decay conditions on $|\hat{\phi}(\xi)|$ as $|\xi| \to \infty$.

DEFINITION 5.4 *Let $\alpha = n + \beta$ with $n \in \mathbb{N}_0$, $0 \leq \beta < 1$. The space C^α consists of the n times continuously differentiable functions f whose nth derivative $D^n f$ is Hölder continuous of order β, that is,*

$$|D^n f(x) - D^n f(y)| \leq c|x - y|^\beta$$

for some constant c.

It is known that if

$$\int (1 + |\xi|)^\alpha |\hat{\phi}(\xi)|\, d\xi < \infty,$$

then $\phi \in C^\alpha$. If we can show that

$$|\hat{\phi}(\xi)| \leq c|\xi|^d$$

Chapter 5: Existence and Regularity

for any $d < -1$, then $\phi \in C^\alpha$ for $0 < \alpha < -d - 1$. ϕ will be continuous, and compact support makes it an L^1-function.

Theorem 5.1 shows that $|\hat{\phi}(\xi)|$ grows no faster than $|\xi|^d$ for $d = \log_2 \sup |h(\xi)|$, but that is not sufficient: since $h(0) = 1$, d cannot be negative. To ensure decay of $\hat{\phi}$, we need approximation order conditions.

THEOREM 5.5
Assume that $h(0) = 1$ and that h satisfies the sum rules of order p, so that

$$h(\xi) = \left(\frac{1 + e^{-i\xi}}{2}\right)^p h_0(\xi)$$

with $h_0(0) = 1$. If

$$\sup_\xi |h_0(\xi)| < 2^{p-\alpha-1},$$

then $\phi \in C^\alpha$. In particular, ϕ is n times continuously differentiable, where n is the largest integer $\leq \alpha$.

PROOF It suffices to show that

$$|\hat{\phi}(\xi)| \leq c(1 + |\xi|)^{-\alpha - 1 - \epsilon} \tag{5.5}$$

for some $\epsilon > 0$.

We define

$$\Pi_{0,n}(\xi) = \prod_{k=1}^n h_0(2^{-k}\xi) \tag{5.6}$$

in analogy with equation (5.3).

As before,

$$\hat{\phi}(\xi) = \prod_{k=1}^\infty h(2^{-k}\xi)\,\hat{\phi}(0)$$

$$= \prod_{k=1}^\infty \left(\frac{1 + e^{-i2^{-k}\xi}}{2}\right)^p \Pi_{0,\infty}(\xi)\,\hat{\phi}(0).$$

The first product can be evaluated in closed form as

$$\prod_{k=1}^\infty \left(\frac{1 + e^{-i2^{-k}\xi}}{2}\right)^p = \left(\frac{1 - e^{-i\xi}}{i\xi}\right)^p \leq c_1(1 + |\xi|)^{-p}.$$

The second product can be estimated as in equation (5.4). If

$$\sup_\xi |h_0(\xi)| = 2^{p-\alpha-1-\epsilon},$$

then for large enough n,

$$|\Pi_{0,n}(\xi)| \leq |\xi|^{p-\alpha-1-\epsilon}e^c \leq c_2 \cdot (1+|\xi|)^{p-\alpha-1-\epsilon}.$$

Together, this gives equation (5.5). ∎

As a minor extension one can replace the bound on $|h_0|$ in theorem 5.5 by

$$\sup_{\xi} |\Pi_{0,\ell}(\xi)|^{1/n} < 2^{p-\alpha-1}$$

for some n.

Example 5.1
For the scaling function of the Daubechies wavelet D_2, the symbol factors as

$$h(\xi) = \left(\frac{1+e^{-i\xi}}{2}\right)^2 \left(\frac{1+\sqrt{3}}{2} + \frac{1-\sqrt{3}}{2}e^{-i\xi}\right).$$

Thus, $p = 2$ and

$$h_0(\xi) = \frac{1+\sqrt{3}}{2} + \frac{1-\sqrt{3}}{2}e^{-i\xi}.$$

$|h_0(\xi)|$ has a maximum of $\sqrt{3}$. Theorem 5.5 shows that $\phi \in C^\alpha$ for $\alpha < 1 - \log_2 \sqrt{3} \approx 0.2075$. This implies that $\phi(x)$ is continuous. □

5.3 L^2-Theory

In this section we will look for sufficient conditions for the existence of an L^2-solution of the refinement equation. This will lead to further smoothness estimates for the solution, generally better than the L^1-estimates.

5.3.1   Transition Operator

DEFINITION 5.6   *The* transition operator *or* transfer operator *for the symbol $h(\xi)$ is defined by*

$$Tf(\xi) = |h(\tfrac{\xi}{2})|^2 f(\tfrac{\xi}{2}) + |h(\tfrac{\xi}{2}+\pi)|^2 f(\tfrac{\xi}{2}+\pi). \tag{5.7}$$

The transition operator maps 2π-periodic functions into 2π-periodic functions, but it also maps some smaller spaces into themselves.

Chapter 5: Existence and Regularity

Let E_n be the space of trigonometric polynomials of the form

$$f(\xi) = \sum_{k=-n}^{n} f_k e^{-ik\xi},$$

and let F_n be the subspace of those $f \in E_n$ with $f(0) = 0$.

LEMMA 5.7
If $h(\xi)$ has degree N, the transition operator T maps E_n into itself for any $n \geq N - 1$.
If $h(\pi) = 0$, then T also maps F_n into itself for any $n \geq N - 1$.

PROOF For trigonometric polynomials, the operation

$$f(\xi) \to f(\frac{\xi}{2}) + f(\frac{\xi}{2} + \pi)$$

corresponds to downsampling:

$$\sum_k f_k e^{-ik\xi} \to 2 \sum_k f_{2k} e^{-ik\xi}.$$

The transition operator is a multiplication by $|h(\xi)|^2$, followed by downsampling.

$|h(\xi)|^2$ has exponents ranging from $-N$ to N. For $f \in E_n$, $|h(\xi)|^2 f(\xi)$ will have exponents ranging from $-N - n$ to $N + n$. The downsampling will remove the odd numbers in that range, and cut the even numbers in half; Tf lies in E_r, where r is the largest integer $\leq (N + n)/2$. If $n \geq N - 1$, then $r \leq n$.

If $h(\pi) = 0$ and $f \in F_n$, $n \geq N - 1$, then Tf lies in E_n and

$$Tf(0) = |h(0)|^2 f(0) + |h(\pi)|^2 f(\pi) = 0;$$

thus, Tf lies in F_n. ∎

We denote the restriction of T to E_n by T_n.
The following lemma is the main trick which will allow us to get better decay estimates for $|\hat{\phi}(\xi)|$.

LEMMA 5.8
If T is the transition operator for $h(\xi)$, then for any $f \in L^2[-\pi, \pi]$

$$\int_{-2^k \pi}^{2^k \pi} f(2^{-k}\xi) |\Pi_k(\xi)|^2 \, d\xi = \int_{-\pi}^{\pi} T^k f(\xi) \, d\xi.$$

This is proved in [39, lemma 3.2], using induction on k.

Where does the transition operator come from? If $\phi \in L^2$, we can define the function

$$a(x) = \langle \phi(y), \phi(y-x) \rangle = \int \phi(y) \phi(y-x)^* \, dy.$$

Its Fourier transform is given by

$$\hat{a}(\xi) = \sqrt{2\pi} \, |\hat{\phi}(\xi)|^2,$$

which makes it useful for studying the L^2-properties of ϕ. However, this is not quite the function we want to use.

DEFINITION 5.9 *The* autocorrelation function *of $\phi \in L^2$ is defined as*

$$\omega(\xi) = \sum_k a(k) e^{-ik\xi}. \tag{5.8}$$

If ϕ has compact support, only finitely many of the $a(k)$ will be nonzero, so ω is a trigonometric polynomial. It is easier to compute and easier to work with than $\hat{a}(\xi)$.

By using the Poisson summation formula, we can verify that

$$\omega(\xi) = \sqrt{2\pi} \sum_k |\hat{\phi}(\xi + 2\pi k)|^2;$$

thus,

$$\int_0^{2\pi} \omega(\xi) \, d\xi = \sqrt{2\pi} \int_{\mathbb{R}} |\hat{\phi}(\xi)|^2 \, d\xi.$$

In the cascade algorithm, let $\omega^{(n)}$ be the autocorrelation function of $\phi^{(n)}$. We can then verify that

$$\omega^{(n+1)} = T\omega^{(n)},$$

so the properties of the transition operator are intimately related to the convergence of $\omega^{(n)}$, which in turn is related to the L^2-convergence of $\phi^{(n)}$.

For starters, we see that T must have an eigenvalue of 1, with the autocorrelation function as the corresponding eigenfunction.

For practical computations, we can work with matrices instead of operators.

DEFINITION 5.10 *The* transition matrix *or* transfer matrix T_n *is defined by*

$$T_{k\ell} = \sum_s h_s h^*_{s+\ell-2k}, \qquad -n \leq k, \ell \leq n. \tag{5.9}$$

Chapter 5: Existence and Regularity

There is a close relationship between the transition matrix and the transition operator, which is why we use the same notation for both.

We identify the function $f = \sum_k f_k e^{-ik\xi} \in E_n$ with the vector $\mathbf{f} \in \mathbb{C}^{2n+1}$ with the same coefficients. The vector and the function have equivalent norms:

$$\int_{-\pi}^{\pi} |f(\xi)|^2 \, d\xi = 2\pi \|\mathbf{f}\|_2^2.$$

For $n \geq N - 1$, the function $T_n f \in E_n$ (where T_n is the transition operator) then corresponds to the vector $T_n \mathbf{f} \in \mathbb{C}^{2n+1}$ (where T_n is the transition matrix). We can switch back and forth between the two viewpoints. In particular, this gives us an easy way to compute eigenvalues and eigenvectors of T_n.

5.3.2 Sobolev Space Estimates

DEFINITION 5.11 *The function f lies in the* Sobolev space H_s *if*

$$\int (1 + |\xi|)^{2s} |\hat{f}(\xi)|^2 \, d\xi < \infty.$$

Sobolev spaces are nested: $H_s \supset H_t$ if $s \leq t$. H_0 is L^2. If $f \in H_s$ for any $s \geq 0$, then $f \in L^2$.

Also, $f \in H_s$ implies $f \in C^\alpha$ for any $\alpha < s - 1/2$, as the following estimate shows:

$$\int (1 + |\xi|)^\alpha |\hat{f}(\xi)| \, d\xi = \int (1 + |\xi|)^{-1/2-\epsilon} (1 + |\xi|)^{\alpha+1/2+\epsilon} |\hat{f}(\xi)| \, d\xi$$

$$\leq \left(\int (1 + |\xi|)^{-1-2\epsilon} \, d\xi \right)^{1/2} \left(\int (1 + |\xi|)^{2\alpha+1+2\epsilon} |\hat{f}(\xi)|^2 \, d\xi \right)^{1/2}.$$

THEOREM 5.12
Assume that $h(0) = 1$ and that h satisfies the sum rules of order p, so that

$$h(\xi) = \left(\frac{1 + e^{-i\xi}}{2} \right)^p h_0(\xi)$$

with $h_0(0) = 1$.

Choose a trigonometric polynomial γ with the following properties:

- $\gamma(\xi) \geq 0$ *for all ξ.*

- $\gamma(\xi) \geq c > 0$ *for $\pi/2 \leq |\xi| \leq \pi$.*

Let ρ be the spectral radius of the transition operator T for h_0, restricted to the smallest invariant subspace that contains the functions $T^k\gamma$, $k \geq 0$. Then $\phi \in H_s$ for any $s < p - \log_4 \rho$.

PROOF The proof given here is based on [41]. A similar approach is used in [65].

We begin the same way as in theorem 5.5. For $|\xi| \leq 2^N\pi$,

$$|\hat{\phi}(\xi)|^2 = \left|\prod_{k=1}^{\infty} \frac{1 + e^{-i2^{-k}\xi}}{2}\right|^{2p} \left|\prod_{k=1}^{N} h_0(2^{-k}\xi)\right|^2 \left|\prod_{k=N+1}^{\infty} h_0(2^{-k}\xi)\right|^2$$
$$\leq c_1(1 + |\xi|)^{-2p} |\Pi_{0,N}(\xi)|^2 e^{2c_2}.$$

At this point, we could estimate the remaining product as in theorem 5.5. This would prove that $\phi \in H_s$ for any $s < \alpha + 1/2$ (same α as in theorem 5.5), but we can do better than that. The key is lemma 5.8.

We find that

$$\left(\int_{-2^N\pi}^{-2^{N-1}\pi} + \int_{2^{N-1}\pi}^{2^N\pi}\right) |\Pi_{0,N}(\xi)|\, d\xi$$
$$\leq \frac{1}{c}\left(\int_{-2^N\pi}^{-2^{N-1}\pi} + \int_{2^{N-1}\pi}^{2^N\pi}\right) \gamma(2^{-N}\xi)|\Pi_{0,N}(\xi)|\, d\xi$$
$$\leq \frac{1}{c}\int_{-2^N\pi}^{2^N\pi} \gamma(2^{-N}\xi)|\Pi_{0,N}(\xi)|\, d\xi$$
$$= \frac{1}{c}\int_{-\pi}^{\pi} T^N\gamma(\xi)\, d\xi$$
$$\leq c_2(\rho + \epsilon)^N$$

for any $\epsilon > 0$. The constant c_2 depends on γ and ϵ, but not on N. In the last step, we have used the fact that if ρ is the spectral radius of T, then for any $\epsilon > 0$ we can find a norm so that

$$\|Tf\| \leq (\rho + \epsilon)\|f\|.$$

Then

$$\left(\int_{-2^N\pi}^{-2^{N-1}\pi} + \int_{2^{N-1}\pi}^{2^N\pi}\right)(1 + |\xi|)^{2s}|\hat{\phi}(\xi)|^2\, d\xi$$
$$\leq c_3\, 2^{2Ns}\left(\int_{-2^N\pi}^{-2^{N-1}\pi} + \int_{2^{N-1}\pi}^{2^N\pi}\right)|\hat{\phi}(\xi)|^2\, d\xi \qquad (5.10)$$
$$\leq c_4\, 2^{2Ns} 2^{-2Np}(\rho + \epsilon)^N = c_4\left[4^{s-p}(\rho + \epsilon)\right]^N.$$

We already know that $\hat{\phi}(\xi)$ is bounded on $[-\pi, \pi]$. Adding the pieces for $N \geq 1$, we obtain a finite integral on $(-\infty, \infty)$ if

$$|4^{s-p}(\rho + \epsilon)| < 1,$$

or $s < p - \log_4 \rho$. ∎

REMARK 5.13 The function γ should be chosen so that the subspace generated by its iterates is as small as possible.

The usual choices are $\gamma(\xi) = 1$, which lies in E_0, and $\gamma(\xi) = 1 - \cos \xi$, which lies in F_1. Results from [92] imply that the optimal choice is $\gamma(\xi) = (1 - \cos \xi)^{2p}$.

I do not know of any example where the choice of γ makes a difference in practice. ∎

Example 5.2
For the scaling function of the Daubechies wavelet D_2, the approximation order is $p = 2$ and

$$h_0(\xi) = \frac{1 + \sqrt{3}}{2} + \frac{1 - \sqrt{3}}{2} e^{-i\xi}.$$

Its transition operator leaves the space E_0 invariant. If we use $\gamma(\xi) = 1$, we get $T_0 = 4$ (a 1×1 matrix), which leads to $s = 1$; this in turn implies that $\phi \in C^\alpha$ for any $\alpha < 1/2$. That is a better result than the L^1-estimate in example 5.1. ∎

Both theorem 5.5 and theorem 5.12 rely on pulling out a factor related to the approximation order. However, the effect of this differs.

In the L^1-theory, the factorization is crucial. Since $h_0(0) = 1$, the supremum of $|h_0|$ is always at least 1. We would never get a useful estimate without the contribution of approximation order. We could do a partial factorization of the symbol, but the best result comes from taking p to be the full approximation order.

In the L^2-theory, the factorization lets us work with a smaller transition matrix, which makes life easier, but it has no effect on the final result. That is a consequence of the following theorem.

THEOREM 5.14
Assume $h(\xi)$ is a scaling function symbol of degree n, and

$$h_1(\xi) = \left(\frac{1 + e^{-i\xi}}{2}\right) h(\xi).$$

Let T_{n-1} be the transition matrix of h on E_{n-1}, and T_n the transition matrix of h_1 on E_n.

If the eigenvalues of T_{n-1} are λ_k, $k = 1, \ldots, 2n - 1$, then the eigenvalues of T_n are 1, $1/2$, and $\lambda_k/4$, $k = 1, \ldots, 2n - 1$.

The proof can be found in [134].

This theorem says the following: if ϕ has approximation order p, its transition matrix will automatically have eigenvalues $1, 1/2, 1/4, \ldots, (1/2)^{2p-1}$. Let λ be the magnitude of the largest of the remaining eigenvalues.

If we compute the spectral radius of T for the full symbol with a suitable γ, the invariant subspace generated by iterates of γ will not include the eigenvectors to the power-of-two eigenvalues. We will find $\rho = \lambda$ (or possibly an even smaller number).

If we factor out the approximation orders first, the power-of-two eigenvalues will disappear. We will find $\rho = 4^p \lambda$, but $s = p - \log_4 \rho$ will be unchanged.

This gives us a very easy L^2-estimate: determine the approximation order and the eigenvalues of the transition matrix. Remove the known power-of-two eigenvalues. The largest remaining eigenvalue is an upper bound on ρ, which leads to a lower bound on s. The bound may not be optimal, but it is easy to find.

DEFINITION 5.15 *A matrix A satisfies* Condition E *if it has a single eigenvalue of 1, and all other eigenvalues are smaller than 1 in absolute value.*

More generally, A satisfies Condition E(p) *if it has a p-fold nondegenerate eigenvalue of 1, and all other eigenvalues are smaller than 1 in absolute value.*

A p-fold eigenvalue is nondegenerate if it has p linearly independent eigenvectors.

THEOREM 5.16
Let n be the degree of $h(\xi)$.
A sufficient condition for $\phi \in L^2$ is that $h(\pi) = 0$ and that the transition matrix T_{n-1} satisfies condition E.

PROOF From the definition of the transition matrix T_n we see that every column sums to either

$$\sum_k h_k \sum_k h^*_{2k} \quad \text{or} \quad \sum_k h_k \sum_k h^*_{2k+1}.$$

If $h(\pi) = 0$, all the column sums are 1. This means that $\mathbf{e}^* = (1, 1, \ldots, 1)$ is a left eigenvector of T_{n-1} to eigenvalue 1. The space F_{n-1} is the orthogonal complement of $\mathbf{e}$, so it consists precisely of the eigenspaces for all the other

Chapter 5: Existence and Regularity

eigenvalues of T_{n-1}. By condition E, the spectral radius of T restricted to F_{n-1} is less than 1.

Now we use theorem 5.12 without factoring out the approximation orders: $p = 0$, $\rho < 1$, so $\phi \in H_s$ for some $s > 0$. ∎

5.3.3 Cascade Algorithm

One way of obtaining a solution to the refinement equation is to use fixed point iteration on it. If the iteration converges in L^2, this presents a practical way of approximating the function $\phi(x)$, as well as an existence proof for the solution. This was already mentioned in chapter 1.

DEFINITION 5.17 *Assume we are given $h(\xi)$ with $h(0) = 1$. The cascade algorithm consists of selecting a suitable starting function $\phi^{(0)}(x) \in L^2$, and then producing a sequence of functions*

$$\phi^{(n+1)}(x) = \sqrt{2} \sum_{k=k_0}^{k_1} h_k \phi^{(n)}(2x - k),$$

or equivalently

$$\hat{\phi}^{(n+1)}(\xi) = h(\xi/2)\hat{\phi}^{(n)}(\xi/2) = \Pi_n(\xi)\hat{\phi}^{(0)}(2^{-n-1}\xi).$$

THEOREM 5.18
Assume that $h(\xi)$ satisfies $h(0) = 1$ and $h(\pi) = 0$. If the transition operator T satisfies condition E and the starting function $\phi^{(0)}$ satisfies

$$\sum_k \phi^{(0)}(x - k) = c \neq 0,$$

then the cascade algorithm converges in L^2.

PROOF This is shown for

$$\phi^{(0)}(x) = \sqrt{2\pi} \frac{\sin \pi x}{\pi x}$$

in [39], and for $\phi^{(0)}$ = Haar function in [134]. The general proof is given in [131]. ∎

The following theorem was already quoted in theorems 1.3 and 1.18.

THEOREM 5.19
If the cascade algorithm converges for both ϕ and $\tilde{\phi}$, and the symbols $h(\xi)$ and $\tilde{h}(\xi)$ satisfy the biorthogonality conditions in equation (1.15), then ϕ and ψ are biorthogonal.

The proof can be found in [39]. The basic idea is this: we start with $\phi^{(0)}$, $\tilde{\phi}^{(0)}$ which are biorthogonal. Conditions in equation (1.15) will ensure that biorthogonality is preserved for all pairs $\phi^{(n)}$, $\tilde{\phi}^{(n)}$, and convergence of the cascade algorithm will ensure that it is preserved in the limit.

5.4 Pointwise Theory

In section 1.8, we already explained how to compute point values of ϕ at the integers, and via repeated application of the refinement relation at all dyadic points.

A dyadic point is a rational number of the form

$$x = \frac{k}{2^n}, \quad n \in \mathbb{N}, \quad k \in \mathbb{Z}.$$

The following describes a more formalized way of doing this, which can then be used to obtain smoothness estimates.

To keep the notation simpler, assume that $\operatorname{supp} \phi = [0, n]$. We define

$$\boldsymbol{\phi}(x) = \begin{pmatrix} \phi(x) \\ \phi(x+1) \\ \vdots \\ \phi(x+n-1) \end{pmatrix}, \quad x \in [0,1]. \tag{5.11}$$

This is related to the constant vector $\boldsymbol{\phi}$ in section 1.8, but it is not quite the same: $\boldsymbol{\phi}(0)$ consists of the first n values of the previous $\boldsymbol{\phi}$, $\boldsymbol{\phi}(1)$ of the last n values.

The recursion relation states that

$$\phi(x+k) = \sqrt{2} \sum_\ell h_\ell \phi(2x + 2k - \ell) = \sqrt{2} \sum_\ell h_{2k-\ell} \phi(2x + \ell).$$

Obviously,
$$\phi(x+k) = [\boldsymbol{\phi}(x)]_k$$
(the kth entry in the vector $\boldsymbol{\phi}(x)$), while

$$\phi(2x+\ell) = \begin{cases} [\boldsymbol{\phi}(2x)]_\ell & \text{if } 0 \le x \le 1/2, \\ [\boldsymbol{\phi}(2x-1)]_{\ell+1} & \text{if } 1/2 \le x \le 1. \end{cases}$$

If $0 \le x \le 1/2$, then

$$[\boldsymbol{\phi}(x)]_k = \sqrt{2} \sum_\ell h_{2k-\ell} [\boldsymbol{\phi}(2x)]_\ell,$$

Chapter 5: Existence and Regularity

or
$$\phi(x) = T_0\phi(2x), \tag{5.12}$$
where
$$(T_0)_{k\ell} = \sqrt{2}h_{2k-\ell}, \quad 0 \le k,\ell \le n-1.$$
If $1/2 \le x \le 1$, then
$$[\phi(x)]_k = \sqrt{2}\sum_\ell h_{2k-\ell+1}[\phi(2x-1)]_\ell,$$
or
$$\phi(x) = T_1\phi(2x-1), \tag{5.13}$$
where
$$(T_1)_{k\ell} = \sqrt{2}h_{2k-\ell+1} \quad 0 \le k,\ell \le n-1.$$

The matrices T_0 and T_1 are related to the matrix T from section 1.8: $T(0)$ is T with the last row and column deleted. T_1 is T with the first row and column deleted, and the indices renumbered. If $h(\pi) = 0$, then $\mathbf{e}^* = (1,1,\ldots,1)^*$ is a common left eigenvector of T_0 and T_1 to eigenvalue 1.

Choose a dyadic number x. In binary notation, we can express x as
$$x = (0.d_1d_2\cdots d_k)_2, \quad d_i = 0 \text{ or } 1.$$
Define the shift operator τ by
$$\tau x = (0.d_2d_3\cdots d_k)_2;$$
then
$$\tau x = \begin{cases} 2x & \text{if } 0 \le x \le 1/2, \\ 2x-1 & \text{if } 1/2 \le x \le 1. \end{cases}$$
Equations (5.12) and (5.13) together are equivalent to
$$\phi(x) = T_{d_1}\phi(\tau x),$$
or after repeated application
$$\phi(x) = T_{d_1}\cdots T_{d_k}\phi(0). \tag{5.14}$$

Example 5.3
Suppose we are interested in the value of the scaling function of the Daubechies wavelet D_2 at the point $3/4$.

T_0 and T_1 are the top left and bottom right 3×3 submatrices of the matrix T from example 1.9, so
$$T_0 = \sqrt{2}\begin{pmatrix} h_0 & 0 & 0 \\ h_2 & h_1 & h_0 \\ 0 & h_3 & h_2 \end{pmatrix}, \quad T_1 = \sqrt{2}\begin{pmatrix} h_1 & h_0 & 0 \\ h_3 & h_2 & h_1 \\ 0 & 0 & h_3 \end{pmatrix}.$$

The vector $\phi(0)$ is

$$\phi(0) = \begin{pmatrix} 0 \\ (1+\sqrt{3})/2 \\ (1-\sqrt{3})/2 \end{pmatrix}.$$

For $x = 3/4 = (0.11)_2$,

$$\phi(3/4) = \begin{pmatrix} \phi(3/4) \\ \phi(7/4) \\ \phi(11/4) \end{pmatrix} = T_1 T_1 \phi(0) = \begin{pmatrix} (9+5\sqrt{3})/16 \\ (2-2\sqrt{3})/16 \\ (5-3\sqrt{3})/16 \end{pmatrix}. \qquad \square$$

The same approach can be used to find point values of the wavelet function. We define matrices S_0, S_1 analogous to T_0, T_1

$$(S_0)_{k\ell} = \sqrt{2} g_{2k-\ell}, \quad (S_1)_{k\ell} = \sqrt{2} g_{2k-\ell+1}, \quad 0 \le k, \ell \le n-1,$$

and replace the leftmost matrix in equation (5.14):

$$\psi(x) = S_{d_1} T_{d_2} \cdots T_{d_k} \phi(0).$$

Example 5.4
We take again the scaling function of the Daubechies wavelet D_2. The refinement matrices for the wavelet function are

$$S_0 = \sqrt{2} \begin{pmatrix} g_0 & 0 & 0 \\ g_2 & g_1 & g_0 \\ 0 & g_3 & g_2 \end{pmatrix} = \sqrt{2} \begin{pmatrix} h_3 & 0 & 0 \\ h_1 & -h_2 & h_3 \\ 0 & -h_0 & h_1 \end{pmatrix},$$

and similarly for S_1.
For $x = 0$,

$$\psi(0) = \begin{pmatrix} \psi(0) \\ \psi(1) \\ \psi(2) \end{pmatrix} = S_0 \phi(0) = \begin{pmatrix} 0 \\ (1-\sqrt{3})/2 \\ (-1-\sqrt{3})/2 \end{pmatrix}.$$

For $x = 3/4 = (0.11)_2$,

$$\psi(3/4) = \begin{pmatrix} \psi(3/4) \\ \psi(7/4) \\ \psi(11/4) \end{pmatrix} = S_1 T_1 \phi(0) = \begin{pmatrix} (-3-\sqrt{3})/16 \\ (-14+2\sqrt{3})/16 \\ (1-\sqrt{3})/16 \end{pmatrix}. \qquad \square$$

This approach can be extended to find exact point values of refinable functions at any rational point, but we are interested in smoothness estimates.

DEFINITION 5.20 *The (uniform) joint spectral radius of T_0, T_1 is defined as*

$$\rho(T_0, T_1) = \limsup_{\ell \to \infty} \max_{d_k = 0 \text{ or } 1} \|T_{d_1} \cdots T_{d_\ell}\|^{1/\ell}.$$

THEOREM 5.21
Assume $h(\pi) = 0$ and that

$$\rho(T_0|F_1, T_1|F_1) = \lambda < 1,$$

where F_1 is the orthogonal complement of the common left eigenvector **e** of T_0, T_1.

Then the refinement equation has a unique solution ϕ which is Hölder continuous of order α for any

$$\alpha < -\log_2 \lambda.$$

PROOF First, we show that ϕ exists and is continuous.

Fix some α with $0 < \alpha < -\log_2 \lambda$ and choose $\epsilon > 0$ so that $\lambda + \epsilon = 2^{-\alpha} < 1$.

From the definition of the joint spectral radius, we can find an $N \in \mathbb{N}$ so that on F_1,

$$\|T_{d_1} T_{d_2} \cdots T_{d_\ell}\| \leq (\lambda + \epsilon)^\ell$$

for $\ell > N$ and all choices of d_k.

Determine the values of ϕ at the integers, normalized so that $\sum_k \phi(k) = 1$, as in section 1.8. Let $\phi^{(0)}$ be the piecewise linear function which interpolates at these points, and let $\phi^{(0)}$ be defined as in equation (5.11). Then

$$\mathbf{e}^T \phi^{(0)}(x) = 1 \qquad \text{for all } x.$$

Let

$$\phi^{(n+1)}(x) = T_{d_1} \phi^{(n)}(\tau x).$$

(This is the cascade algorithm starting with $\phi^{(0)}$.) Since **e** is a left eigenvector of both T_0 and T_1, we have $\mathbf{e}^* \phi^{(n)}(x) = 1$ for all n and x. The difference between any two $\phi^{(n)}(x)$ and $\phi^{(m)}(x)$ lies in F_1.

Then for $n > N$,

$$\begin{aligned}\|\phi^{(n+1)}(x) - \phi^{(n)}(x)\| &= \|T_{d_1} T_{d_2} \cdots T_{d_n}(\phi^{(1)}(\tau^n x) - \phi^{(0)}(\tau^n x))\| \\ &\leq c(\lambda + \epsilon)^n,\end{aligned} \quad (5.15)$$

where

$$c = \sup_{x \in [0,1]} \|\phi^{(1)}(x) - \phi^{(0)}(x)\|.$$

For $\ell > n > N$, we find

$$\|\phi^{(\ell)}(x) - \phi^{(n)}(x)\| \leq \sum_{k=n}^{\ell-1} \|\phi^{(k+1)}(x) - \phi^{(k)}(x)\| < c \frac{(\lambda + \epsilon)^n}{1 - (\lambda + \epsilon)}.$$

This shows that $\{\phi^{(n)}(x)\}$ is a Cauchy sequence uniformly in x; it converges to a continuous limit function $\phi(x)$, which gives us ϕ.

The Hölder continuity is proved in a similar manner. Assume $x, y \in [0, 1]$ match in the first n binary digits but not in digit $n + 1$, so $2^{-n-1} < |x - y| \leq 2^{-n}$. Then

$$|\phi(x) - \phi(y)| \leq \|\phi(x) - \phi(y)\| = \|T_{d_1} T_{d_2} \cdots T_{d_n}(\phi(\tau^n x) - \phi(\tau^n y))\|$$
$$\leq c(\lambda + \epsilon)^n = c2^{-\alpha n} \leq 2^\alpha c |x - y|^\alpha.$$

There is still a bit more to do here: if $|x - y| \leq 2^{-n}$, they do not necessarily have to match in the first n digits. Details can be found in [52]. ∎

It is shown in [43] that this estimate cannot be improved: ϕ is not Hölder continuous of any order larger than $-\log_2 \lambda$. The estimate may or may not hold for $\alpha = -\log_2 \lambda$.

Generalizations of theorem 5.21 can be used to guarantee higher orders of smoothness, both globally and on certain subsets of the support of ϕ.

THEOREM 5.22

Assume that $\phi(x)$ has approximation order p. Let F_p be the orthogonal complement of $\text{span}\{\mathbf{e}_0, \ldots, \mathbf{e}_{p-1}\}$, where

$$\mathbf{e}_k^* = (0^k, 1^k, 2^k, \ldots),$$

and

$$\rho(T_0|F_p, T_1|F_p) = \lambda < 1.$$

Then $\phi \in C^\alpha$ for any $\alpha < -\log_2 \lambda$.

Example 5.5

For the scaling function of the Daubechies wavelet D_2, T_0 and T_1 are of size 3×3. Since $p = 2$, F_2 is one-dimensional. In fact, $T_0|_{F_2} = (1 + \sqrt{3})/4$, $T_1|_{F_2} = (1 - \sqrt{3})/4$. We can calculate the exact $\lambda = (1 + \sqrt{3})/4$, so $\phi \in C^{0.55}$.

This is a better estimate than in examples 5.1 and 5.2, and it is the best possible. ∎

This was a very easy example. In general, a joint spectral radius can be quite hard to compute. With some effort, it is possible to get reasonably good estimates.

From the definition we get the estimate

$$\rho(T_0, T_1) \leq \max(\|T_0\|, \|T_1\|)$$

or more generally

$$\rho(T_0, T_1) \leq \max_{d_k = 0 \text{ or } 1} \|T_{d_1} \cdots T_{d_\ell}\|^{1/\ell}$$

for any fixed ℓ. Moderately large ℓ give good bounds in practice, but the number of norms we have to examine grows like 2^ℓ.

We can also get lower bounds on ρ this way, which puts upper bounds on the smoothness. For any specific choice of d_k,

$$\rho(T_0, T_1) \geq \|T_{d_1} \cdots T_{d_\ell}\|^{1/\ell}.$$

The frequency domain methods can only give lower bounds on the smoothness estimates. The pointwise method can give both upper and lower bounds.

As with the L^1- and L^2-estimates, it is a good idea to take advantage of the approximation order p. This improves the pointwise estimates by taking the joint spectral radius on a smaller subspace F_p.

5.5 Smoothness and Approximation Order

We saw in the preceding sections that approximation order is important for smoothness estimates.

A high approximation order does not directly guarantee smoothness, but the two tend to be correlated. For some families of scaling functions with increasing approximation order, a corresponding increase in smoothness can be proved.

For example, the Sobolev exponent of the Daubechies wavelet D_p increases by approximately $1 - \log_4 3 \approx 0.2075$ for every increase in p (see [49] and [65].) The Sobolev exponent of the smoothest possible orthogonal scaling function of given length increases asymptotically at a slightly faster rate [147].

It can also be shown that smoothness implies a certain minimum approximation order.

THEOREM 5.23
If ϕ, ψ and $\tilde{\phi}$, $\tilde{\psi}$ form biorthogonal wavelets of compact support, and if ϕ is p times continuously differentiable, then $\tilde{\psi}$ has at least $(p+1)$ vanishing moments.

PROOF This is a brief explanation of the basic idea of the proof. The complete proof (with more general conditions that do not require compact support) can be found in [50, section 5.5].

$\psi(x)$ and $\tilde{\psi}(x)$ are biorthogonal, which means

$$\langle \psi_{mk}(x), \tilde{\psi}_{n\ell}(x) \rangle = \delta_{mn}\delta_{k\ell}.$$

We do induction on p. Assume we have already shown that

$$\int x^k \tilde{\psi}(x)\, dx = 0 \quad \text{for } k = 0, \ldots, p-1.$$

Since $\phi(x)$ is p times continuously differentiable, so is $\psi(x)$. Find a dyadic point $x_0 = k/2^m$ somewhere where $D^p \psi(x_0) \neq 0$; then ψ_{mk} will have a nonzero pth derivative at 0.

If we approximate ψ_{mk} by its Taylor polynomial t_p of order p around zero, t_p will have a nonzero leading coefficient.

For large n, $\tilde{\psi}_{n0}$ is concentrated in a very small interval around 0, and

$$0 = \langle \psi_{mk}, \tilde{\psi}_{n0} \rangle \approx \langle t_p, \tilde{\phi}_{n0} \rangle \approx c \langle x^p, \tilde{\phi}_{n0} \rangle.$$

∎

5.6 Stability

The previous sections discussed the existence of $\phi(x)$ and its smoothness properties. In order to ensure that ϕ produces an MRA we need to also verify that

- ϕ has stable shifts.
- $\bigcap_k V_k = \{0\}$.
- $\overline{\bigcup_k V_k} = L^2$.

We will now give sufficient conditions for these properties.

Recall that ϕ has stable shifts if $\phi \in L^2$ and if there exist constants $0 < A \leq B$ so that for all sequences $\{c_k\} \in \ell_2$,

$$A \sum_k |c_k|^2 \leq \| \sum_k c_k^* \phi(x-k) \|_2^2 \leq B \sum_k |c_k|^2.$$

LEMMA 5.24
$\phi \in L^2$ has stable shifts if and only if there exist constants $0 < A \leq B$ so that

$$A \leq \sum_k |\hat{\phi}(\xi + 2k\pi)|^2 \leq B.$$

PROOF The "if" part is easy. For given $\{c_k\} \in \ell_2$, let

$$c(\xi) = \sum_k c_k e^{-ik\xi};$$

Chapter 5: Existence and Regularity   117

then
$$\|c\|_2^2 = \int_0^{2\pi} |c(\xi)|^2 \, d\xi = 2\pi \sum_k |c_k|^2.$$

The Fourier transform of $\sum_k c_k^* \phi(x-k)$ is $c(\xi)^* \hat{\phi}(\xi)$, so

$$\|\sum_k c_k^* \phi(x-k)\|_2^2 = \|c(\xi)^* \hat{\phi}(\xi)\|_2^2$$

$$= \int |c(\xi)|^2 |\hat{\phi}(\xi)|^2 \, d\xi$$

$$= \int_0^{2\pi} |c(\xi)|^2 \sum_k |\hat{\phi}(\xi + 2\pi k)|^2 \, d\xi.$$

If
$$A \le \sum_k |\hat{\phi}(\xi + 2k\pi)|^2 \le B,$$

then
$$2\pi A \sum_k |c_k|^2 \le \|\sum_k c_k^* \phi(x-k)\|_2^2 \le 2\pi B \sum_k |c_k|^2.$$

For the "only if" part we have to work a little harder. The 2π-periodic function $\sum_k |\hat{\phi}(\xi+2k\pi)|^2$ is continuous, since it is a trigonometric polynomial with coefficients $\langle \phi(x), \phi(x-k) \rangle$. If it vanishes for some ξ_0, then for any $\epsilon > 0$ we can choose a neighborhood of ξ_0 where $|\phi(\xi)| < \epsilon$, and a $c(\xi)$ with support inside this neighborhood. Then

$$\|\sum_k c_k^* \phi(x-k)\|_2^2 \le \epsilon \sum_k |c_k|^2,$$

which contradicts the existence of a lower bound A. ∎

LEMMA 5.25
If $\phi, \tilde{\phi} \in L^2$ are biorthogonal, they have stable shifts.

PROOF   This proof comes from [39, page 319].
We define the *correlation function* the same way as the autocorrelation function (definition 5.9), but with

$$a(x) = \langle \phi(y), \tilde{\phi}(y-x) \rangle = \int \phi(y) \tilde{\phi}(y-x)^* \, dy.$$

Then
$$\omega(\xi) = \sqrt{2\pi} \sum_k \hat{\phi}(\xi + 2\pi k) \widehat{\tilde{\phi}}(\xi + 2\pi k)^*.$$

Biorthogonality implies that $\omega(\xi) \equiv 1$, which leads to

$$\left(\sum_k |\hat{\phi}(\xi + 2\pi k)|^2\right)\left(\sum_k |\hat{\tilde{\phi}}(\xi + 2\pi k)|^2\right) \geq \frac{1}{\sqrt{2\pi}}.$$

This implies nonzero lower bounds A, $\tilde{A}$ in lemma 5.24. The upper bounds are simply

$$B = 2\pi\|\phi\|_2^2, \qquad \tilde{B} = 2\pi\|\tilde{\phi}\|_2^2. \qquad \blacksquare$$

A concept related to stability is *linear independence*. A compactly supported scaling function ϕ has *linearly independent shifts* if

$$\sum_j \mathbf{a}_j^* \phi(x - j) = 0 \Rightarrow \mathbf{a} = 0 \tag{5.16}$$

for all sequences $\mathbf{a}$.

It is shown in [88] and [89] that ϕ has linearly independent shifts if and only if the sequences

$$\{\hat{\phi}(\xi + 2\pi k)\}_{k \in \mathbb{Z}} \tag{5.17}$$

are linearly independent for all $\xi \in \mathbb{C}$. ϕ has stable shifts if and only if the sequences in equation (5.17) are linearly independent for all $\xi \in \mathbb{R}$.

Thus, linear independence implies stability.

THEOREM 5.26

(i) If ϕ is a refinable L^2-function with stable shifts, then

$$\bigcap_k V_k = \{0\}.$$

(ii) If ϕ is a refinable L^2-function with stable shifts, $\hat{\phi}$ bounded, continuous at 0 and with $\hat{\phi}(0) \neq 0$, then

$$\overline{\bigcup_k V_k} = L^2.$$

The proof is given in [50], section 5.3.2.

All the conditions for a pair of biorthogonal MRAs are satisfied if ϕ, $\tilde{\phi}$ are compactly supported, biorthogonal L^2-functions.

For a full justification of decomposition and reconstruction, we need to show that the wavelet functions are also stable.

There are actually two definitions of stability that we need to consider. Stability at a single level means that there exist constants $0 < A \leq B$ so that

$$A \sum_k |c_k|^2 \leq \|\sum_k c_k^* \psi(x - k)\|_2^2 \leq B \sum_k |c_k|^2.$$

This condition is automatically satisfied if there exists a pair of biorthogonal wavelet functions. That is lemma 5.25. Stability at level 0 implies stability at any other fixed level n, which is already enough to justify a decomposition and reconstruction over a finite number of levels.

Stability over all levels means

$$A \sum_{nk} |c_{nk}|^2 \leq \| \sum_{nk} c_{nk}^* 2^{n/2} \psi(2^n x - k)\|_2^2 \leq B \sum_{nk} |c_{nk}|^2.$$

This is required if we want to decompose an L^2-function f in terms of wavelet functions over all levels.

THEOREM 5.27
A pair of biorthogonal symbols h, $\tilde{h}$ with $h(0) = \tilde{h}(0) = 1$, $h(\pi) = \tilde{h}(\pi) = 0$ generates biorthogonal bases of compactly supported wavelet functions, stable over all levels, if and only if the transition operators T and $\tilde{T}$ both satisfy condition E.

The proof is rather lengthy. It can be found in [39, theorem 5.1].

Part II
Multiwavelets

6

Basic Theory

This chapter introduces the basic concepts of multiwavelet theory. It runs in parallel with the classical (scalar) wavelet theory in chapter 2. Readers with sufficient background in scalar wavelets can skip the first part of the book and begin here.

The multiwavelet theory presented in part II of this book is more general than the scalar wavelet theory in part I in two respects.

First, the main difference is of course the switch from scalar wavelets to multiwavelets. The recursion coefficients are now matrices, the symbols are trigonometric matrix polynomials, and so on. This change is responsible for most of the extra complication.

Second, we now consider a dilation factor of m rather than 2. This can also be done for scalar wavelets. It adds a little bit of complexity, but not much, and results can be stated in greater generality. Mostly, it complicates the notation. With a dilation factor of $m > 2$ we still have one scaling function, but we get $m - 1$ wavelets instead of one.

The standard notation for the case $m = 2$ is to use ϕ, ψ for the scaling and wavelet function, h_k and g_k for their recursion coefficients, etc. For general m, it is common to use $\phi^{(0)}$ for the scaling function, $\phi^{(1)}, \ldots, \phi^{(m-1)}$ for the wavelets, and similarly for the symbols and other quantities.

This notation has the advantage that results can be stated more concisely, but that same conciseness also makes it harder on the reader. It is easier to read and understand two formulas, one for the scaling function and one for the wavelet functions, than a single formula that covers both, usually with a Kronecker delta somewhere.

I have chosen to continue using different letters. The multiscaling function is still $\boldsymbol{\phi}$ (now in boldface), the multiwavelets are $\boldsymbol{\psi}^{(1)}, \ldots, \boldsymbol{\psi}^{(m-1)}$. Likewise, the recursion coefficients are H_k and $G_k^{(1)}, \ldots, G_k^{(m-1)}$ (now in uppercase), and so on.

6.1 Refinable Function Vectors

DEFINITION 6.1 *A refinable function vector is a vector-valued function*

$$\phi(x) = \begin{pmatrix} \phi_1(x) \\ \vdots \\ \phi_r(x) \end{pmatrix}, \qquad \phi_n : \mathbb{R} \to \mathbb{C},$$

which satisfies a two-scale matrix refinement equation *of the form*

$$\phi(x) = \sqrt{m} \sum_{k=k_0}^{k_1} H_k\, \phi(mx - k), \quad k \in \mathbb{Z}. \tag{6.1}$$

r is called the multiplicity *of ϕ; the integer $m \geq 2$ is the* dilation factor. *The recursion coefficients H_k are $r \times r$ matrices.*

The refinable function vector ϕ is called orthogonal *if*

$$\langle \phi(x), \phi(x-k) \rangle = \int \phi(x)\phi(x-k)^* \, dx = \delta_{0k} I.$$

This inner product is an $r \times r$ matrix. Throughout this book, I always stands for an identity matrix of the appropriate size.

It is possible to consider matrix refinement equations with an infinite sequence of recursion coefficients. Most of multiwavelet theory remains valid in this case, as long as the coefficients decay rapidly enough. Still, allowing infinite sequences of recursion coefficients requires additional technical conditions in many theorems, and complicates the proof. We will always assume that there are only finitely many nonzero recursion coefficients, which covers most cases of practical interest.

Example 6.1

A simple example with multiplicity 2 and dilation factor 2 is the *constant/linear* refinable function vector (fig. 6.1)

$$\phi(x) = \begin{pmatrix} 1 \\ \sqrt{3}(2x - 1) \end{pmatrix}, \qquad x \in [0, 1]. \tag{6.2}$$

It satisfies

$$\phi_1(x) = \phi_1(2x) + \phi_1(2x+1),$$

$$\phi_2(x) = \left[-\frac{\sqrt{3}}{2}\phi_1(2x) + \frac{1}{2}\phi_2(2x) \right] + \left[\frac{\sqrt{3}}{2}\phi_1(2x+1) + \frac{1}{2}\phi_2(2x+1) \right].$$

Chapter 6: Basic Theory

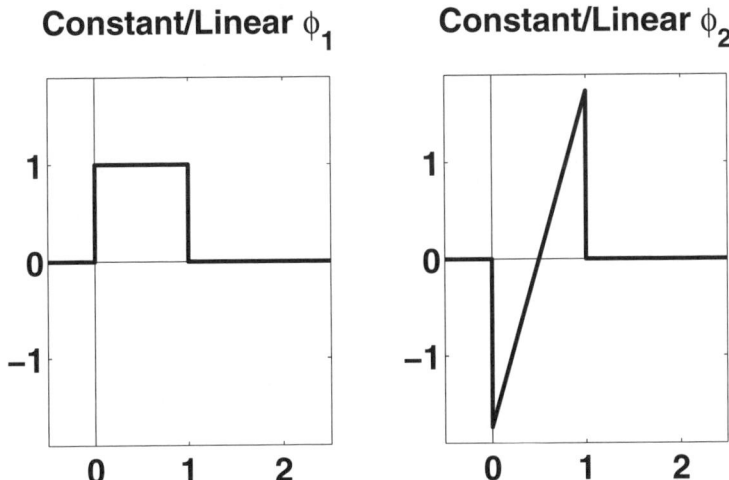

FIGURE 6.1
Two components of constant/linear refinable function vector.

This function vector is refinable with

$$H_0 = \frac{1}{2\sqrt{2}} \begin{pmatrix} 2 & 0 \\ -\sqrt{3} & 1 \end{pmatrix}, \qquad H_1 = \frac{1}{2\sqrt{2}} \begin{pmatrix} 2 & 0 \\ \sqrt{3} & 1 \end{pmatrix}.$$

It is orthogonal, which is easy to check directly from equation (6.2). □

One can generalize this example to multiplicity r by using the powers of x up to x^{r-1}, orthonormalized on $[0,1]$. In fact, that leads to the original construction of Alpert [2].

Example 6.2

Another common example, also with multiplicity 2 and dilation factor 2, is the *Hermite cubic* refinable function vector (fig. 6.2)

$$\phi_1(x) = \begin{cases} 3x^2 - 2x^3 & \text{for } x \in [0,1], \\ 3(2-x)^2 - 2(2-x)^3 & \text{for } x \in [1,2], \end{cases}$$
$$\phi_2(x) = \begin{cases} x^3 - x^2 & \text{for } x \in [0,1], \\ (2-x)^2 - (2-x)^3 & \text{for } x \in [1,2]. \end{cases} \qquad (6.3)$$

These functions are cubic splines with one continuous derivative and sup-

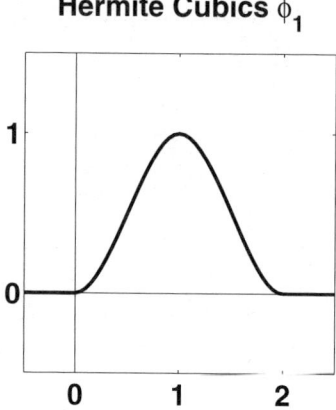

FIGURE 6.2
The two components of the Hermite cubic refinable function vector.

port $[0, 2]$ which satisfy

$$\phi_1(1) = 1, \quad \phi_1'(1) = 0,$$
$$\phi_2(1) = 0, \quad \phi_2'(1) = 1.$$

This function vector is refinable with

$$H_0 = \frac{1}{8\sqrt{2}} \begin{pmatrix} 4 & 6 \\ -1 & -1 \end{pmatrix}, \quad H_1 = \frac{1}{8\sqrt{2}} \begin{pmatrix} 8 & 0 \\ 0 & 4 \end{pmatrix}, \quad H_2 = \frac{1}{8\sqrt{2}} \begin{pmatrix} 4 & -6 \\ 1 & -1 \end{pmatrix}.$$

It is not orthogonal. ☐

Example 6.3
The *Donovan–Geronimo–Hardin–Massopust (DGHM)* multiscaling function [59] is commonly considered to be the first nontrivial example of a multiscaling function (fig. 6.3. It has recursion coefficients

$$H_0 = \frac{1}{20\sqrt{2}} \begin{pmatrix} 12 & 16\sqrt{2} \\ -\sqrt{2} & -6 \end{pmatrix}, \quad H_1 = \frac{1}{20\sqrt{2}} \begin{pmatrix} 12 & 0 \\ 9\sqrt{2} & 20 \end{pmatrix},$$
$$H_2 = \frac{1}{20\sqrt{2}} \begin{pmatrix} 0 & 0 \\ 9\sqrt{2} & -6 \end{pmatrix}, \quad H_3 = \frac{1}{20\sqrt{2}} \begin{pmatrix} 0 & 0 \\ -\sqrt{2} & 0 \end{pmatrix}.$$

It is orthogonal. ☐

Many refinable functions are well-defined but cannot be written in closed form; the DGHM function is one example. Nevertheless, we can compute val-

Chapter 6: Basic Theory

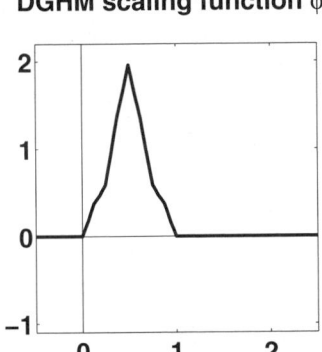
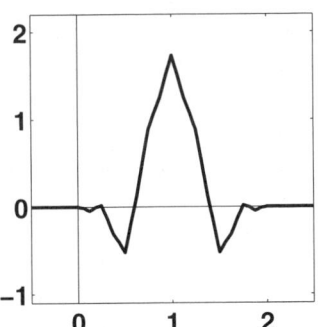

FIGURE 6.3
The two components of the DGHM refinable function vector.

ues at individual points (or at least approximate them to arbitrary accuracy), compute integrals, determine smoothness properties, and more.

The details will be presented in later chapters. Right now, we just briefly mention two of these techniques, to introduce concepts we need in this chapter.

DEFINITION 6.2 *The symbol of a refinable function vector is the trigonometric matrix polynomial*

$$H(\xi) = \frac{1}{\sqrt{m}} \sum_{k=k_0}^{k_1} H_k e^{-ik\xi}.$$

The Fourier transform of the refinement equation is

$$\hat{\phi}(\xi) = H(\xi/m)\hat{\phi}(\xi/m). \tag{6.4}$$

By substituting this relation into itself repeatedly and taking the limit, we find that formally

$$\hat{\phi}(\xi) = \left[\prod_{k=1}^{\infty} H(m^{-k}\xi)\right] \hat{\phi}(0). \tag{6.5}$$

The infinite product is a product of matrices, so we have to be careful about the order. In the expanded product, k increases from left to right.

Assuming that the infinite product converges, this provides a way to compute $\phi(x)$, at least in principle.

Choosing $\hat{\phi}(0) = \mathbf{0}$ gives $\phi = \mathbf{0}$, which is not an interesting solution, so we want $\hat{\phi}(0) \neq \mathbf{0}$.

The vector $\hat{\phi}(0)$ is not arbitrary: assuming that $\hat{\phi}$ is continuous at 0, equation (6.4) implies that a nontrivial solution can only exist if $H(0)$ has an eigenvalue of 1, and $\hat{\phi}(0)$ is a corresponding eigenvector.

The *length* of this eigenvector is still arbitrary. Solutions of refinement equations are only defined up to constant factors: any multiple of a solution is also a solution.

The infinite product approach is useful for existence and smoothness estimates (see chapter 11), but it is not a practical way of finding ϕ. A way to get approximate point values of $\phi(x)$ is the *cascade algorithm*, which is fixed point iteration applied to the refinement equation.

We choose a suitable starting function $\phi^{(0)}$, and define

$$\phi^{(n)}(x) = \sqrt{m} \sum_k H_k \phi^{(n-1)}(mx - k).$$

This will converge in many cases.

Orthogonality of a refinable function can be checked directly from the recursion coefficients or the symbol.

THEOREM 6.3
A necessary condition for orthogonality is

$$\sum_k H_k H_{k-m\ell}^* = \delta_{0\ell} I, \tag{6.6}$$

or equivalently

$$\sum_{n=0}^{m-1} |H(\xi + \frac{2\pi n}{m})|^2 = I. \tag{6.7}$$

These conditions are sufficient to ensure orthogonality if the cascade algorithm for ϕ converges.

PROOF Substitute the refinement equation into the definition of orthogonality:

$$\begin{aligned}
\delta_{0\ell} I &= \langle \phi(x), \phi(x-\ell) \rangle \\
&= m \sum_{k,n} H_k H_n^* \langle \phi(mx-k), \phi(mx - m\ell - n) \rangle \\
&= \sum_{k,n} H_k H_n^* \langle \phi(y), \phi(y + k - m\ell - n) \rangle \\
&= \sum_{k,n} H_k H_n^* \delta_{0, k-m\ell-n} = \sum_k H_k H_{k-m\ell}^*.
\end{aligned}$$

In terms of the symbol,

$$\sum_{n=0}^{m-1}|H(\xi+\frac{2\pi n}{m})|^2 = \frac{1}{m}\sum_{n=0}^{m-1}\sum_{k,s}H_k H_s^* e^{-i(k-s)\xi}e^{-in(k-s)2\pi/m}$$
$$= \sum_{k,\ell}H_k H_{k-m\ell}^* e^{-im\ell\xi} = I, \qquad (6.8)$$

where we have used

$$\sum_{n=0}^{m-1}e^{-in(k-s)2\pi/m} = \begin{cases} m & \text{if } k-s=m\ell, \quad \ell \in \mathbb{N}, \\ 0 & \text{otherwise.} \end{cases}$$

The sufficiency can be proved with the same argument as in the scalar case: we start the cascade algorithm with an orthogonal function $\phi^{(0)}$. The orthogonality relations make sure that orthogonality is preserved at every step. If the cascade algorithm converges, the limit function must be orthogonal. ∎

As an example, we can use equation (6.6) to verify that the DGHM multi-scaling function is orthogonal.

In the scalar case, the orthogonality relations imply that an orthogonal refinable function has to have an even number of recursion coefficients, since otherwise one of the orthogonality relations is $h_{k_0}h_{k_1}^* = 0$.

In the multiwavelet case, there is no restriction: a product of two nonzero matrices can be zero.

DEFINITION 6.4 *The* support *of a function ϕ is defined as the closure of the set*

$$\{x : \phi(x) \neq 0\}.$$

The support of a function vector $\boldsymbol{\phi}$ is defined as

$$\text{supp } \boldsymbol{\phi} = \bigcup_k \text{supp } \phi_k.$$

"Compact support" means the same as "bounded support."

THEOREM 6.5
If $\boldsymbol{\phi}$ is a solution of equation (6.1) with compact support, then

$$\text{supp } \boldsymbol{\phi} \subset \left[\frac{k_0}{m-1}, \frac{k_1}{m-1}\right].$$

PROOF Assume supp $\phi = [a, b]$. When we substitute this into the refinement equation, we find that

$$\text{supp } \phi \subset \left[\frac{a + k_0}{m}, \frac{b + k_1}{m}\right].$$

This implies that

$$a \geq \frac{a + k_0}{m}, \qquad b \leq \frac{b + k_1}{m},$$

or

$$a \geq \frac{k_0}{m - 1}, \qquad b \leq \frac{k_1}{m - 1}. \qquad \blacksquare$$

The same argument also shows that if we start the cascade algorithm with a function $\phi^{(0)}$ with support $[a^{(0)}, b^{(0)}]$, the support of $\phi^{(n)}$ will converge to some subset of $[k_0/(m-1), k_1/(m-1)]$ as $n \to \infty$.

In the scalar case, the support of ϕ was precisely equal to the bounds. This is not necessarily the case for multiwavelets.

Some components of ϕ could have a smaller support than ϕ as a whole. If those components are the only ones used near the endpoints in the refinement equation, then ϕ could have a support strictly contained in the interval $[k_0/(m-1), k_0/(m-1)]$. For example, the DGHM refinable function vector has support $[0, 2]$ instead of the expected $[0, 3]$.

It is shown in [111] that under some minor technical conditions the support of ϕ is precisely $[k_0/(m-1), k_1/(m-1)]$ provided that the first and last recursion coefficients H_{k_0}, H_{k_1} are not nilpotent. A matrix A is called *nilpotent* if $A^n = 0$ for some n. In the DGHM example, H_3 is nilpotent, which is why the support is shorter on the right.

For practical applications we need ϕ to have some minimal regularity properties.

DEFINITION 6.6 *The refinable function vector ϕ has* stable shifts *if $\phi \in L^2$ and if there exist constants $0 < A \leq B$ so that for all sequences of vectors $\{\mathbf{c}_k\} \in (\ell_2)^r$,*

$$A \sum_k \|\mathbf{c}_k\|_2^2 \leq \|\sum_k \mathbf{c}_k^* \phi(x - k)\|_2^2 \leq B \sum_k \|\mathbf{c}_k\|_2^2.$$

If ϕ is orthogonal, it is automatically stable with $A = B = 1$. This follows from lemma 11.21.

DEFINITION 6.7 *A compactly supported refinable function vector ϕ has* linearly independent shifts *if for all sequences of vectors $\{\mathbf{c}_k\}$,*

$$\sum_k \mathbf{c}_k^* \phi(x - k) = 0 \Rightarrow \mathbf{c}_k = \mathbf{0} \quad \text{for all } k.$$

There are no conditions on the sequence $\{c_k\}$. Compact support guarantees that for each fixed k the sum only contains finitely many nonzero terms.

It can be shown that linear independence implies stability.

DEFINITION 6.8 *A matrix A satisfies* Condition E *if it has a simple eigenvalue of 1, and all other eigenvalues are smaller than 1 in absolute value.*

More generally, A satisfies Condition E(p) *if it has a p-fold nondegenerate eigenvalue of 1, and all other eigenvalues are smaller than 1 in absolute value.*

A p-fold eigenvalue is nondegenerate if it has p linearly independent eigenvectors.

THEOREM 6.9
Assume that ϕ is a compactly supported L^2-solution of the refinement equations with nonzero integral and linearly independent shifts. This implies the following conditions:

(i) $H(0)$ satisfies condition E.

(ii) There exists a vector $\mathbf{y}_0 \neq \mathbf{0}$ so that

$$\sum_k \mathbf{y}_0^* \phi(x-k) = c, \qquad c \text{ constant.}$$

(iii) The same vector $\mathbf{y}_0$ satisfies

$$\mathbf{y}_0^* H\left(\frac{2\pi k}{m}\right) = \delta_{0k} \mathbf{y}_0^*, \qquad k = 0, \ldots, m-1.$$

(iv) The same vector $\mathbf{y}_0$ satisfies

$$\mathbf{y}_0^* \sum_\ell H_{m\ell+k} = \frac{1}{\sqrt{m}} \mathbf{y}_0^*, \qquad k = 0, \ldots, m-1.$$

The proof can be found in [122].

We will always assume from now on that ϕ satisfies the properties listed in theorem 6.9. We will refer to them as the *basic regularity conditions*.

6.2 MRAs and Multiwavelets

The definitions and explanations in this section are essentially the same as in sections 1.2 and 1.4, with suitable minor modifications. It is expected that the reader is already familiar with the concept of multiresolution approximations (MRAs) and wavelets, so the discussion here is briefer than in the first part of the book. If necessary, refer back to the earlier sections.

6.2.1 Orthogonal MRAs and Multiwavelets

DEFINITION 6.10 *An MRA of L^2 is a doubly infinite nested sequence of subspaces of L^2*

$$\cdots \subset V_{-1} \subset V_0 \subset V_1 \subset V_2 \subset \cdots$$

with properties

(i) $\bigcup_n V_n$ *is dense in* L^2.

(ii) $\bigcap_n V_n = \{0\}$.

(iii) $f(x) \in V_n \iff f(mx) \in V_{n+1}$ *for all* $n \in \mathbb{Z}$.

(iv) $f(x) \in V_n \iff f(x - m^{-n}k) \in V_n$ *for all* $n, k \in \mathbb{Z}$.

(v) *There exists a function vector* $\phi \in L^2$ *so that*

$$\{\phi_\ell(x - k) : \ell = 1, \ldots, r,\ k \in \mathbb{Z}\} \tag{6.9}$$

forms a stable basis of V_0.

The vector of basis functions ϕ is called the multiscaling *function.*
The MRA is called orthogonal *if ϕ is orthogonal.*

Condition (v) means that any $f \in V_0$ can be written uniquely as

$$f(x) = \sum_{k \in \mathbb{Z}} \mathbf{f}_k^* \, \phi(x - k)$$

with convergence in the L^2-sense; and there exist constants $0 < A \leq B$, independent of f, so that

$$A \sum_k \|\mathbf{f}_k\|_2^2 \leq \|f\|_2^2 \leq B \sum_k \|\mathbf{f}_k\|_2^2.$$

This implies that ϕ has stable shifts.

Chapter 6: Basic Theory

Condition (iii) expresses the main property of an MRA: each V_n consists of the functions in V_0 compressed by a factor of m^n. Thus, a stable basis of V_n is given by $\{\phi_{nk} : k \in \mathbb{Z}\}$, where

$$\phi_{nk}(x) = m^{n/2}\phi(m^n x - k). \tag{6.10}$$

The factor $m^{n/2}$ preserves the L^2-norm.

Since $V_0 \subset V_1$, ϕ can be written in terms of the basis of V_1 as

$$\phi(x) = \sum_k H_k \phi_{1k}(x) = \sqrt{m} \sum_k H_k \phi(mx - k)$$

for some coefficient matrices H_k. In other words, ϕ is refinable (with possibly an infinite sequence of coefficients). We will assume that the refinement equation is in fact a finite sum.

Let us assume for now that the MRA is orthogonal. In the scalar case, this implies that ϕ is unique (modulo a shift and scaling by a constant of magnitude 1). This is not true in the multiwavelet case.

DEFINITION 6.11 *A trigonometric matrix polynomial $A(\xi)$ is paraunitary if*

$$A(\xi)A(\xi)^* = A(\xi)^*A(\xi) = I.$$

This generalizes the definition of a unitary matrix

$$AA^* = A^*A = I.$$

THEOREM 6.12
Assume that ϕ_1 and ϕ_2 are orthogonal multiscaling functions with compact support. They span the same space V_0 if and only if

$$\phi_2(x) = \sum_k A_k \phi_1(x - k), \tag{6.11}$$

where

$$A(\xi) = \sum_k A_k e^{-ik\xi}$$

is a paraunitary matrix polynomial.

PROOF If ϕ_2 spans the same space as ϕ_1, the two function vectors must be related by equation (6.11) for some finite sequence A_k. The orthogonality relations for ϕ_1 and ϕ_2 imply that

$$\sum_k A_k A_{k-\ell}^* = \delta_{0\ell} I,$$

which is equivalent to the fact that $A(\xi)$ is paraunitary.

Conversely, if $A(\xi)$ is paraunitary and we define ϕ_2 by equation (6.11), then also
$$\phi_1(x) = \sum_k A_k^* \phi_2(x+k),$$
and they span the same space. ∎

As a special case, if
$$\phi_2 = Q\phi_1$$
for refinable ϕ_1 and unitary Q, then ϕ_1 and ϕ_2 span the same space V_0, and their recursion coefficients are related by conjugation with Q. If
$$\phi_1(\xi) = \sqrt{m} \sum_k H_k \phi_1(mx-k),$$
then
$$\phi_2(\xi) = \sqrt{m} \sum_k (QH_kQ^*) \phi_2(mx-k).$$
This is a special case of a regular two-scale similarity transform (TST), defined in chapter 8.

REMARK 6.13 For biorthogonal multiwavelets, A just needs to be invertible, rather than paraunitary. You may find that some multiwavelets are listed with different recursion coefficients in other papers. That is usually due to a simple transformation with diagonal A, that is, a rescaling of the functions. ∎

The orthogonal projection of an arbitrary function $f \in L^2$ onto V_n is given by
$$P_n f(x) = \sum_k \langle f, \phi_{nk} \rangle \phi_{nk}(x).$$
As in the scalar case, the basis functions ϕ_{nk} are shifted in steps of m^{-n}; so $P_n f$ is interpreted as an approximation to f at *resolution* m^{-n} or *scale* m^{-n}. As $n \to \infty$, $P_n f$ converges to f in the L^2-sense. This can be shown as in the scalar case.

The difference between the approximations at resolution m^{-n} and m^{-n-1} is the fine detail at resolution m^{-n}:
$$Q_n f(x) = P_{n+1} f(x) - P_n f(x).$$
Q_n is also an orthogonal projection. Its range W_n is orthogonal to V_n, and
$$V_n \oplus W_n = V_{n+1}.$$

Chapter 6: Basic Theory

The two sequences of spaces $\{V_n\}$ and $\{W_n\}$ and their relationships can be graphically represented as in figure 1.4.

As in the scalar case,

$$V_n = \bigoplus_{k=-\infty}^{n-1} W_k,$$

and the projections $P_n f$ converge to f in L^2 as $n \to \infty$.

The sequence of spaces $\{W_n\}$ satisfies conditions similar to those of an MRA. The symbol $\perp$ stands for "is orthogonal to."

THEOREM 6.14
For any orthogonal MRA with multiscaling function ϕ,

(i) $\bigoplus_n W_n$ *is dense in L^2.*

(ii) $W_k \perp W_n$ *if $k \neq n$.*

(iii) $f(x) \in W_n \iff f(mx) \in W_{n+1}$ *for all $n \in \mathbb{Z}$.*

(iv) $f(x) \in W_n \iff f(x - m^{-n}k) \in W_n$ *for all $n, k \in \mathbb{Z}$.*

(v) *There exist function vectors $\boldsymbol{\psi}^{(s)} \in L^2$, $s = 1, \ldots, m-1$, orthogonal to $\boldsymbol{\phi}$ and to each other, so that*

$$\{\boldsymbol{\psi}^{(s)}(x-k) : s = 1, \ldots, m-1, k \in \mathbb{Z}\}$$

forms a stable basis of W_0, and

$$\{\boldsymbol{\psi}^{(s)}_{nk} : s = 1, \ldots, m-1, n, k \in \mathbb{Z}\}$$

forms a stable basis of L^2.

(vi) *Since each $\boldsymbol{\psi}^{(s)} \in V_1$, it can be represented as*

$$\boldsymbol{\psi}^{(s)}(x) = \sqrt{m} \sum_k G_k^{(s)} \boldsymbol{\phi}(mx - k)$$

for some coefficients $G_k^{(s)}$.

The function vectors $\boldsymbol{\psi}^{(s)}$ are called the **multiwavelet functions**. $\boldsymbol{\phi}$ and $\boldsymbol{\psi}^{(s)}$ together form a **multiwavelet**.

The proof is essentially the same as in the scalar case (theorem 6), except that we do not have an explicit formula for the recursion coefficients of $\boldsymbol{\psi}$. We will explain in chapter 10 how to find them. There is no simple formula like in the scalar case, and the multiwavelet functions are not unique. The construction will show that it is always possible to find $\boldsymbol{\psi}^{(s)}$ of compact support if $\boldsymbol{\phi}$ has compact support.

The fact that we have $m-1$ multiwavelet functions instead of 1 has nothing to do with multiwavelets. It comes from using a dilation factor m instead of 2.

We define the symbol of $\boldsymbol{\psi}^{(s)}$ as

$$G^{(s)}(\xi) = \frac{1}{\sqrt{m}} \sum_k G_k^{(s)} e^{-ik\xi}.$$

The Fourier transform of the refinement equation for $\boldsymbol{\psi}^{(s)}$ is

$$\hat{\boldsymbol{\psi}}^{(s)}(\xi) = G^{(s)}(\frac{\xi}{m})\hat{\boldsymbol{\phi}}(\frac{\xi}{m}).$$

The orthogonality of $\boldsymbol{\phi}$ and $\boldsymbol{\psi}^{(s)}$ can be expressed as in theorem 6.3:

$$\begin{aligned} \sum H_k H_{k-m\ell}^* &= I, \\ \sum G_k^{(s)} G_{k-m\ell}^{(t)*} &= \delta_{0\ell}\delta_{st} I, \\ \sum H_k G_{k-m\ell}^{(s)*} &= \sum G_k^{(s)} H_{k-m\ell}^* = 0, \end{aligned} \qquad (6.12)$$

or equivalently

$$\begin{aligned} |H(\xi)|^2 + |H(\xi+\pi)|^2 &= I, \\ G^{(s)}(\xi)G^{(t)}(\xi)^* + G^{(s)}(\xi+\pi)G^{(t)}(\xi+\pi)^* &= \delta_{st} I, \\ H(\xi)G^{(s)}(\xi)^* + H(\xi+\pi)G^{(s)}(\xi+\pi)^* &= 0, \\ G^{(s)}(\xi)H(\xi)^* + G^{(s)}(\xi+\pi)H(\xi+\pi)^* &= 0 \end{aligned} \qquad (6.13)$$

for $s = 1, \ldots, m-1$.

In terms of the multiwavelet functions, the projection Q_n is given by

$$Q_n f = \sum_{s=1}^{m-1} \sum_k \langle f, \boldsymbol{\psi}_{nk}^{(s)} \rangle \boldsymbol{\psi}_{nk}^{(s)}.$$

We now come to the main concept we seek: the *discrete multiwavelet transform* (DMWT).

Given a function $f \in L^2$, we can represent it as

$$f = \sum_{k=-\infty}^{\infty} Q_k f$$

(complete decomposition in terms of detail at all levels), or we can start at any level ℓ and use the approximation at resolution $m^{-\ell}$ plus all the detail at finer resolution:

$$f = P_\ell f + \sum_{k=\ell}^{\infty} Q_k f.$$

Chapter 6: Basic Theory

For practical applications, we need to reduce this to a finite sum. We assume that $f \in V_n$ for some $n > \ell$. Then

$$f = P_n f = P_\ell f + \sum_{k=\ell}^{n-1} Q_k f. \tag{6.14}$$

Equation (6.14) describes the DMWT: the original function or signal f gets decomposed into a coarse approximation $P_\ell f$, and fine detail at several resolutions. The decomposition as well as the reconstruction can be performed very efficiently on a computer. Implementation details are presented in section 7.1.

6.2.2 Biorthogonal MRAs and Multiwavelets

As in the scalar case, we can drop the orthogonality requirement.

DEFINITION 6.15 *Two refinable function vectors ϕ, $\tilde{\phi}$ are called biorthogonal if*

$$\langle \phi(x-k), \tilde{\phi}(x-\ell) \rangle = \delta_{k\ell} I.$$

We also call $\tilde{\phi}$ the dual of ϕ.

THEOREM 6.16
A necessary condition for biorthogonality is

$$\sum H_k \tilde{H}^*_{k-m\ell} = \delta_{0\ell} I, \tag{6.15}$$

or equivalently

$$H(\xi)\tilde{H}(\xi)^* + H(\xi+\pi)H(\xi+\pi)^* = I. \tag{6.16}$$

These conditions are sufficient to ensure biorthogonality if the cascade algorithm for both ϕ and $\tilde{\phi}$ converges.

The proof is analogous to that of theorem 6.3.

As in the scalar case, it is possible to orthonormalize an existing multiscaling function with stable shifts, but the resulting new ϕ does not usually have compact support any more. This makes it less desirable for practical applications.

Assume now that we have two MRAs $\{V_n\}$ and $\{\tilde{V}_n\}$, generated by biorthogonal multiscaling functions ϕ and $\tilde{\phi}$. We can complete the construction of multiwavelets as follows.

The projections P_n and $\tilde{P}_n$ from L^2 into V_n, $\tilde{V}_n$, respectively, are given by

$$P_n f = \sum_k \langle f, \tilde{\phi}_{nk} \rangle \phi_{nk},$$

$$\tilde{P}_n f = \sum_k \langle f, \phi_{nk} \rangle \tilde{\phi}_{nk},$$

where ϕ_{nk}, $\tilde{\phi}_{nk}$ are defined as in equation (6.10). These are now oblique (i.e., nonorthogonal) projections.

The projections Q_n, $\tilde{Q}_n$ are defined as before by

$$Q_n f = P_{n+1} f - P_n f,$$
$$\tilde{Q}_n f = \tilde{P}_{n+1} f - \tilde{P}_n f,$$

and their ranges are the spaces W_n, $\tilde{W}_n$.

The space W_n is orthogonal to $\tilde{V}_n$. This is proved as in the scalar case. We still have

$$V_n \oplus W_n = V_{n+1}$$

as a nonorthogonal direct sum.

THEOREM 6.17
Assume that ϕ, $\tilde{\phi} \in L^2$ are multiscaling functions generating biorthogonal MRAs, and that the cascade algorithm converges for both of them. Then

(i) $\bigoplus_n W_n$, $\bigoplus_n \tilde{W}_n$ are dense in L^2.

(ii) $W_k \perp \tilde{W}_n$ if $k \neq n$.

(iii)

$$f(x) \in W_n \iff f(mx) \in W_{n+1} \quad \text{for all } n \in \mathbb{Z},$$
$$f(x) \in \tilde{W}_n \iff f(mx) \in \tilde{W}_{n+1} \quad \text{for all } n \in \mathbb{Z}.$$

(iv)

$$f(x) \in W_n \iff f(x - m^{-n}k) \in W_n \quad \text{for all } n, k \in \mathbb{Z},$$
$$f(x) \in \tilde{W}_n \iff f(x - m^{-n}k) \in \tilde{W}_n \quad \text{for all } n, k \in \mathbb{Z}.$$

(v) There exist biorthogonal functions $\psi^{(s)}, \tilde{\psi}^{(s)} \in L^2$ so that

$$\{\psi(x-k)^{(s)} : s = 1, \ldots, m-1, k \in \mathbb{Z}\}$$

forms a stable basis of W_0,

$$\{\tilde{\psi}^{(s)}(x-k) : s = 1, \ldots, m-1, k \in \mathbb{Z}\}$$

forms a stable basis of $\tilde{W}_0$, and

$$\{\psi_{nk}^{(s)} : s = 1, \ldots, m-1, n, k \in \mathbb{Z}\},$$
$$\{\tilde{\psi}_{nk}^{(s)} : s = 1, \ldots, m-1, n, k \in \mathbb{Z}\}$$

both form a stable basis of L^2.

Chapter 6: Basic Theory

(vi) Since $\psi^{(s)} \in V_1$, $\tilde{\psi}^{(s)} \in \tilde{V}_1$, they can be represented as

$$\psi^{(s)}(x) = \sqrt{m} \sum_k G_k^{(s)} \phi(mx - k),$$

$$\tilde{\psi}^{(s)}(x) = \sqrt{m} \sum_k \tilde{G}_k^{(s)} \tilde{\phi}(mx - k)$$

for some coefficients $G_k^{(s)}$, $\tilde{G}_k^{(s)}$.

The functions $\psi^{(s)}$, $\tilde{\psi}^{(s)}$ are again called the multiwavelet functions or mother multiwavelets.

The proof is the same as that of theorem 6.14, except for the stability part. That will be covered in chapter 11 (lemma 11.21). Methods for finding the recursion coefficients for the multiwavelet functions will be covered in chapter 10.

As in equations (6.6) and (6.7), we can express the biorthogonality conditions as

$$\sum H_k \tilde{H}_{k-m\ell}^* = \delta_{0\ell} I$$
$$\sum G_k^{(s)} \tilde{G}_{k-m\ell}^{(t)*} = \delta_{0\ell} \delta_{st} I, \qquad (6.17)$$
$$\sum H_k G_{k-m\ell}^{(s)*} = \sum G_k^{(s)} H_{k-m\ell}^* = 0,$$

or equivalently

$$\sum_{k=0}^{m-1} H(\xi + \frac{2\pi k}{m}) \tilde{H}(\xi + \frac{2\pi k}{m})^* = I,$$

$$\sum_{k=0}^{m-1} G^{(s)}(\xi + \frac{2\pi k}{m}) \tilde{G}^{(t)}(\xi + \frac{2\pi k}{m})^* = \delta_{st} I,$$

$$\sum_{k=0}^{m-1} H(\xi + \frac{2\pi k}{m}) \tilde{G}^{(s)*}(\xi + \frac{2\pi k}{m}) = 0, \qquad (6.18)$$

$$\sum_{k=0}^{m-1} G^{(s)}(\xi + \frac{2\pi k}{m}) \tilde{H}(\xi + \frac{2\pi k}{m})^* = 0.$$

As before, $P_n f$ is interpreted as an approximation to f at resolution m^{-n}, and $Q_n f$ is the fine detail. If $f \in V_n$, we can do a multiwavelet decomposition

$$f = P_n f = P_\ell f + \sum_{k=\ell}^{n-1} Q_k f.$$

However, we can also do these things on the dual side. $\tilde{P}_n f$ is also an approximation to f at resolution m^{-n}, with fine detail $\tilde{Q}_n f$. If $f \in \tilde{V}_n$, we can do a dual multiwavelet decomposition

$$f = \tilde{P}_n f = \tilde{P}_\ell f + \sum_{k=\ell}^{n-1} \tilde{Q}_k f.$$

As in the scalar case, formulas for biorthogonal multiwavelets are symmetric: whenever we have any formula or algorithm, we can put a tilde on everything that did not have one, and vice versa, and we get a dual formula or algorithm.

That does not mean that both algorithms are equally useful for a particular application, but they both work.

6.3 Moments

We assume in this section that ϕ satisfies the minimal regularity assumptions from theorem 6.9. In particular, $H(0)$ satisfies condition E.

DEFINITION 6.18 *The kth discrete moments of ϕ, $\psi^{(s)}$ are defined by*

$$M_k = \frac{1}{\sqrt{m}} \sum_\ell \ell^n H_\ell,$$

$$N_k^{(s)} = \frac{1}{\sqrt{m}} \sum_\ell \ell^n G_\ell^{(s)}, \quad s = 1, \ldots, m-1.$$

Discrete moments are $r \times r$ matrices. They are related to the symbols by

$$\begin{aligned} M_k &= i^k D^k H(0), \\ N_k^{(s)} &= i^k D^k G^{(s)}(0). \end{aligned} \quad (6.19)$$

In particular, $M_0 = H(0)$.

Discrete moments are uniquely defined and easy to calculate.

DEFINITION 6.19 *The kth continuous moments of ϕ, $\psi^{(s)}$ are*

$$\begin{aligned} \mu_k &= \int x^k \phi(x)\, dx, \\ \nu_k^{(s)} &= \int x^k \psi^{(s)}(x)\, dx. \end{aligned} \quad (6.20)$$

Chapter 6: Basic Theory

Continuous moments are r-vectors. They are related to the Fourier transforms of ϕ, $\psi^{(s)}$ by

$$\boldsymbol{\mu}_k = \sqrt{2\pi}\, i^k D^k \hat{\boldsymbol{\phi}}(0),$$
$$\boldsymbol{\nu}_k^{(s)} = \sqrt{2\pi}\, i^k D^k \hat{\boldsymbol{\psi}}^{(s)}(0). \tag{6.21}$$

The continuous moment $\boldsymbol{\mu}_0$ is only defined up to a constant multiple by the refinement equation, depending on the scaling of ϕ. For a single ϕ we could pick the factor arbitrarily, but for a biorthogonal pair the normalizations have to match.

LEMMA 6.20
If ϕ, $\tilde{\phi} \in L^1 \cap L^2$ are biorthogonal and satisfy the basic regularity conditions from theorem 6.9, then
$$\tilde{\boldsymbol{\mu}}_0^* \boldsymbol{\mu}_0 = 1.$$

The proof is identical to the scalar case (lemma 1.25).

In the biorthogonal case, we cannot in general achieve both $\|\boldsymbol{\mu}_0\| = 1$ and $\|\tilde{\boldsymbol{\mu}}_0\| = 1$. In the orthogonal case the normalization $\|\boldsymbol{\mu}_0\| = 1$ is mandatory.

THEOREM 6.21
The continuous and discrete moments are related by

$$\boldsymbol{\mu}_k = m^{-k} \sum_{t=0}^{k} \binom{k}{t} M_{k-t} \boldsymbol{\mu}_t,$$
$$\boldsymbol{\nu}_k^{(s)} = m^{-k} \sum_{t=0}^{k} \binom{k}{t} N_{k-t}^{(s)} \boldsymbol{\mu}_t. \tag{6.22}$$

In particular,
$$\boldsymbol{\mu}_0 = M_0 \boldsymbol{\mu}_0 = H(0) \boldsymbol{\mu}_0.$$

Once $\boldsymbol{\mu}_0$ has been chosen, all other continuous moments are uniquely defined and can be computed from these relations.

PROOF We start with
$$\hat{\phi}(m\xi) = H(\xi)\hat{\phi}(\xi)$$

and differentiate k times:
$$m^k \left(D^k \hat{\phi}\right)(m\xi) = \sum_{t=0}^{k} \binom{k}{t} D^{k-t} H(\xi) D^t \hat{\phi}(\xi).$$

When we set $\xi = 0$ and use equations (6.19) and (6.21), we get the first formula in equation (6.22). The second formula is proved similarly.

For $k = 0$ we get
$$\boldsymbol{\mu}_0 = M_0 \boldsymbol{\mu}_0.$$
$\boldsymbol{\mu}_0$ is uniquely defined up to scaling, by condition E.

For $n \geq 1$, equation (6.22) leads to
$$(m^n I - M_0)\boldsymbol{\mu}_n = \sum_{t=0}^{n-1} \binom{n}{t} M_{n-t}\boldsymbol{\mu}_t.$$

Condition E implies that the matrix on the left is nonsingular, so we can compute $\boldsymbol{\mu}_1, \boldsymbol{\mu}_2, \ldots$ successively and uniquely from this. The second formula in equation (6.22) provides the $\boldsymbol{\nu}_n^{(s)}$. ∎

Example 6.4
The first three discrete moments of the DGHM multiscaling function are
$$M_0 = \frac{1}{5}\begin{pmatrix} 3 & 2\sqrt{2} \\ 2\sqrt{2} & 1 \end{pmatrix}, \quad M_1 = \frac{1}{10}\begin{pmatrix} 3 & 0 \\ 6\sqrt{2} & 2 \end{pmatrix}, \quad M_2 = \frac{1}{10}\begin{pmatrix} 3 & 0 \\ 9\sqrt{2} & -1 \end{pmatrix}.$$

With the correct normalization $\|\boldsymbol{\mu}_0\| = 1$, the first three continuous moments are
$$\boldsymbol{\mu}_0 = \frac{1}{3}\begin{pmatrix} \sqrt{6} \\ \sqrt{3} \end{pmatrix}, \quad \boldsymbol{\mu}_1 = \frac{1}{3}\begin{pmatrix} 3\sqrt{6} \\ \sqrt{3} \end{pmatrix}, \quad \boldsymbol{\mu}_2 = \frac{1}{42}\begin{pmatrix} 4\sqrt{6} \\ 13\sqrt{3} \end{pmatrix}.$$
□

6.4 Approximation Order

As pointed out in section 6.2, the projection $P_n f$ of a function f onto the space V_n represents an approximation to f at resolution m^{-n}. How good is this approximation?

DEFINITION 6.22 *The multiscaling function ϕ provides* approximation order p *if*
$$\|f(x) - P_n f(x)\| = O(m^{-np})$$
whenever f has p continuous derivatives.

If ϕ provides approximation order p, then for smooth f
$$\|Q_n f\| = O(m^{-np}),$$

Chapter 6: Basic Theory

since $\|Q_n f\| \leq \|f - P_n f\| + \|f - P_{n+1} f\|$.

DEFINITION 6.23 *The multiscaling function ϕ has accuracy p if all polynomials up to order $p - 1$ can be represented as*

$$x^n = \sum_k \mathbf{c}_{nk}^* \phi(x - k) \tag{6.23}$$

for some coefficient vectors $\mathbf{c}_{nk}$.

This representation is well-defined even though x^n does not lie in the space L_2: Since ϕ has compact support, the sum on the right is finite for any fixed x.

LEMMA 6.24
The coefficients $\mathbf{c}_{nk}$ in equation (6.23) have the form

$$\mathbf{c}_{nk} = \sum_{t=0}^{n} \binom{n}{t} k^{n-t} \mathbf{y}_t,$$

where $\mathbf{y}_t = \mathbf{c}_{t0}$ and $\mathbf{y}_0 \neq \mathbf{0}$.

Here $\mathbf{y}_0$ is the same vector as in theorem 6.9.

PROOF Replace x^n by $(x + k)^n$ in equation (6.23) and expand. ∎

If ϕ has a dual $\tilde{\phi}$, we can multiply equation (6.23) by $\tilde{\phi}(x)$ and integrate to obtain

$$\mathbf{y}_n^* = \langle x^n, \tilde{\phi}(x) \rangle = \tilde{\boldsymbol{\mu}}_n^*. \tag{6.24}$$

DEFINITION 6.25 *The recursion coefficients $\{H_k\}$ of a matrix refinement equation satisfy the* sum rules *of order p if there exist vectors $\mathbf{y}_0, \ldots, \mathbf{y}_{p-1}$ with $\mathbf{y}_0 \neq \mathbf{0}$, which satisfy*

$$\sum_{t=0}^{n} \binom{n}{t} m^t (-i)^{n-t} \mathbf{y}_t^* D^{n-t} H(\frac{2\pi s}{m}) = \begin{cases} \mathbf{y}_n^* & \text{for } s = 0, \\ \mathbf{0}^* & \text{for } s = 1, \ldots, m-1. \end{cases} \tag{6.25}$$

for $n = 0, \ldots, p - 1$.
The vectors $\mathbf{y}_t$ are called the approximation vectors.

The $\mathbf{y}_n$ in lemma 6.24 are the same as the approximation vectors.

Example 6.5
The DGHM multiscaling function satisfies the sum rules of order 2, with approximation vectors

$$\mathbf{y}_0 = \frac{1}{3}\begin{pmatrix}\sqrt{6}\\ \sqrt{3}\end{pmatrix}, \qquad \mathbf{y}_1 = \begin{pmatrix}\frac{1}{6}\sqrt{6}\\ 2\sqrt{3}\end{pmatrix}.$$

LEMMA 6.26
If ϕ satisfies the basic regularity conditions 6.9, then

$$\mathbf{y}_0^* \hat{\phi}(2\pi k) = 0, \quad k \in \mathbb{Z}, k \neq 0.$$

PROOF   Let $k = m^n \ell$, ℓ not divisible by m. Then

$$\begin{aligned}\hat{\phi}(2\pi k) &= H(2\pi m^{n-1}\ell)\hat{\phi}(2\pi m^{n-1}\ell) = \ldots \\ &= H(2\pi m^{n-1}\ell)\cdots H(2\pi\ell)H(\frac{2\pi\ell}{m})\hat{\phi}(\frac{2\pi\ell}{m}).\end{aligned} \quad (6.26)$$

$\mathbf{y}_0$ is a left eigenvector to eigenvalue 1 of $H(0)$; by periodicity of $H(2\pi n)$, $n \in \mathbb{Z}$. $\mathbf{y}_0$ is a left eigenvector of $H(2\pi\ell/m)$ to eigenvalue 0. Thus, if we multiply equation (6.26) from the left by $\mathbf{y}_0^*$, we get 0. ∎

THEOREM 6.27
Assume ϕ is a compactly supported, integrable solution of the matrix refinement equation with linearly independent shifts. Then the following are equivalent:

(i) ϕ has approximation order p.

(ii) ϕ has accuracy p.

(iii) $\{H_k\}$ satisfy the sum rules of order p.

(iv) There exists a trigonometric vector polynomial $\mathbf{a}(\xi)$ with $\mathbf{a}(0) \neq 0$ which satisfies

$$D^n\left[\mathbf{a}^*(m\xi)H(\xi + \frac{2\pi}{m}k)\right]\bigg|_{\xi=0} = \delta_{0k}D^n\mathbf{a}^*(0) \quad (6.27)$$

for $n = 0, \ldots, p-1$.

The approximation vectors $\mathbf{y}_n$ and the function $\mathbf{a}(\xi)$ are related by

$$\mathbf{y}_n = i^n D^n \mathbf{a}(0). \quad (6.28)$$

Chapter 6: Basic Theory 145

(v) *There exists a scalar* superfunction *f which is a finite linear combination of translates of ϕ*

$$f(x) = \sum_k \mathbf{a}_k^* \phi(x-k)$$

and which satisfies the Strang–Fix conditions *of order p, that is*

$$\hat{f}(0) \neq 0,$$
$$\hat{f}^{(n)}(2\pi k) = 0, \quad k \in \mathbb{Z}, \quad k \neq 0, \quad n = 0,\ldots,p-1.$$

If ϕ is part of a biorthogonal multiwavelet, the following condition is also equivalent to the above.

(vi) *The dual multiwavelet functions have p vanishing continuous moments:*

$$\tilde{\boldsymbol{\nu}}_n^{(t)} = \mathbf{0}, \quad t = 1,\ldots,m-1, \quad n = 0,\ldots,p-1.$$

PROOF (i) $\Leftrightarrow$ (ii): This is proved in [87]. It is the most technical part of the proof.

(ii) $\Leftrightarrow$ (iii): We already know that $\{H_k\}$ satisfy the sum rules of order 1 (theorem 6.9).

Assume that ϕ has approximation order p, and we have already established the sum rules of order n for some $n < p$. We write

$$x^n = \sum_k \mathbf{c}_{nk}^* \phi(x-k) = \sqrt{m} \sum_{k\ell} \mathbf{c}_{nk}^* H_\ell \phi(mx - mk - \ell)$$

and compare this to

$$(mx)^n = \sum_k \mathbf{c}_{nk}^* \phi(mx - k).$$

Using the linear independence of translates of ϕ and the lower order sum rules, this gives us some relations for $\{H_k\}$. These can be reduced to the sum rule of order n. The details are lengthy and messy, unfortunately, so we will not present them here.

The details were first published in [80], and independently in [120].

(iii) $\Leftrightarrow$ (iv): We make sure that equation (6.28) holds, by either defining $\mathbf{a}(\xi)$ in terms of $\mathbf{y}_n$ or vice versa.

After that, we just have to expand equation (6.27) using the Leibniz formula (repeated product rule), apply equation (6.28), and we get the sum rules.

(iv) $\Leftrightarrow$ (v): This is shown in [120]. Here is a sketch of the idea: we define f using the coefficients of the same

$$\mathbf{a}(\xi) = \sum_k \mathbf{a}_k e^{ik\xi},$$

as in (iv). Then
$$\hat{f}(\xi) = \mathbf{a}^*(\xi)\hat{\boldsymbol{\phi}}(\xi).$$

This gives us
$$\hat{f}(m\xi) = \mathbf{a}(m\xi)^*\hat{\boldsymbol{\phi}}(m\xi) = \mathbf{a}(m\xi)^*H(\xi)\hat{\boldsymbol{\phi}}(\xi),$$

or if we replace ξ by $(\xi + 2\pi k/m)$,
$$\hat{f}(m\xi + 2\pi k) = \left[\mathbf{a}(m\xi)^*H(\xi + \frac{2\pi}{m}k)\right]\hat{\boldsymbol{\phi}}(\xi + \frac{2\pi}{m}k).$$

Differentiate n times and evaluate at $\xi = 0$ to get
$$m^n D^n \hat{f}(2\pi k) = \sum_{s=0}^{n} \binom{n}{s} D^s \left[\mathbf{a}(m\xi)^*H(\xi + \frac{2\pi}{m}k)\right]\bigg|_{\xi=0} D^{n-s}\hat{\boldsymbol{\phi}}(\frac{2\pi}{m}k)$$
$$= \delta_{0k}\sum_{s=0}^{n}\binom{n}{s}D^s\mathbf{a}(0)^* D^{n-s}\hat{\boldsymbol{\phi}}(\frac{2\pi}{m}k).$$

For $k \neq 0$ everything vanishes. For $k = 0$, $s = 0$ we get
$$\hat{f}(0) = \mathbf{y}_0^*\hat{\boldsymbol{\phi}}(0)$$
$$= \sum_k \mathbf{y}_0^*\hat{\boldsymbol{\phi}}(2\pi k) \qquad (6.29)$$
$$= \frac{1}{2\pi}\sum_k \mathbf{y}_0^*\boldsymbol{\phi}(k) = \frac{1}{2\pi} \neq 0$$

by using lemma 6.26 and the Poisson summation formula.

(iv) $\Leftrightarrow$ (vi): We use the biorthogonality relations in equation (6.18). Multiply the relation
$$\sum_{k=0}^{m-1} H(\xi + \frac{2\pi}{m}k)\tilde{G}^{(s)}(\xi + \frac{2\pi}{m}k)^* = 0$$

from the left by $\mathbf{a}(m\xi)^*$, differentiate n times, and evaluate at $\xi = 0$. The left-hand side becomes
$$\sum_{k=0}^{m-1}\sum_{s=0}^{n}\binom{n}{s}D^s\left[\mathbf{a}(m\xi)^*H(\xi + \frac{2\pi}{m}k)\right]\bigg|_{\xi=0}D^{n-s}\tilde{G}^{(s)}(\xi + \frac{2\pi}{m}k)^*$$
$$= \sum_{s=0}^{n}\binom{n}{s}D^s\mathbf{a}(0)^*D^{n-s}\tilde{G}^{(s)}(0)^*$$
$$= \sum_{s=0}^{n}\binom{n}{s}i^s\tilde{\boldsymbol{\mu}}_s^*i^{n-s}\tilde{N}_{n-s}^* = i^n m^n \tilde{\boldsymbol{\nu}}_n^{(s)*}$$

by equation (6.22).

The reverse direction requires induction on n. ∎

In summary, as in the scalar case, approximation order p, accuracy p, and the sum rules of order p are equivalent for sufficiently regular ϕ. They are also equivalent to the fact that all the dual multiwavelet functions have p vanishing moments, except that we need to specify vanishing *continuous* moments. In the scalar case, vanishing continuous moments are equivalent to vanishing discrete moments, but in the multiwavelet case the discrete moments are matrices. They have to annihilate certain vectors, but they do not have to be zero matrices.

Approximation order p is also equivalent to a certain factorization of the symbol, but not as simple as in the scalar case. This factorization requires a lot of machinery, and will be presented in chapter 8.

REMARK 6.28 At this point in chapter 1, there is a section on symmetry. Symmetry for multiwavelets requires more background, and we delay the discussion until chapter 8.

We just note at this point that there is no restriction against symmetry for multiwavelets. That is one of the properties that make them attractive, compared to scalar wavelets. ∎

6.5 Point Values and Normalization

We mentioned the cascade algorithm earlier in this chapter as a practical way for finding approximate point values of $\phi(x)$. There is another approach but which produces exact point values of ϕ. It usually works, but may fail in some cases.

Recall that the support of ϕ is contained in $[k_0/(m-1), k_1/(m-1)]$. Let a and b be the smallest and largest integers in this interval.

For integer ℓ, $a \leq \ell \leq b$, the refinement equation (6.1) reads

$$\phi(\ell) = \sqrt{m} \sum_{k=k_0}^{k_1} H_k\, \phi(m\ell - k) = \sqrt{m} \sum_{k=k_0}^{k_1} H_{m\ell-k}\, \phi(k).$$

This is an eigenvalue problem

$$\Phi = T\Phi, \qquad (6.30)$$

where

$$\Phi = \begin{pmatrix} \phi(a) \\ \phi(a+1) \\ \vdots \\ \phi(b) \end{pmatrix}, \qquad T_{\ell k} = \sqrt{m}\, H_{m\ell-k}, \quad a \le \ell, k \le b.$$

Note that each column of T contains all of the $H_{mk+\ell}$ for some fixed ℓ. The basic regularity condition (iv) in theorem 6.9 implies that $(\mathbf{y}_0^*, \mathbf{y}_0^*, \ldots, \mathbf{y}_0^*)$ is a left eigenvector to eigenvalue 1, so a right eigenvector also exists.

We assume that this eigenvalue is simple, so that the solution is unique. (This is the place where the algorithm could fail.)

In the case of dilation factor $m = 2$, the support of ϕ is contained in $[a, b] = [k_0, k_1]$. The first and last rows of matrix equation (6.30) are

$$\phi(k_0) = \sqrt{2}\, H_{k_0} \phi(k_0),$$
$$\phi(k_1) = \sqrt{2}\, H_{k_1} \phi(k_1).$$

Unless H_{k_0} or H_{k_1} have an eigenvalue of $1/\sqrt{2}$, the values of ϕ at the endpoints are zero, and we can reduce the size of Φ and T.

Once the values of ϕ at the integers have been determined, we can use the refinement equation to obtain values at points of the form k/m, $k \in \mathbb{Z}$, then k/m^2, and so on to any desired resolution.

All calculations of norms, point values, and moments only give the answer up to an arbitrary constant. This reflects the fact that the refinement equation only defines ϕ up to an arbitrary factor. When we calculate several quantities, how do we make consistent choices? This is called *normalization*.

We already showed that if ϕ satisfies the basic regularity conditions, then

$$\mathbf{y}_0^* \left(\sum_k \phi(k) \right) = \mathbf{y}_0^* \left(\int \phi(x)\, dx \right) = \mathbf{y}_0^* \boldsymbol{\mu}_0 = 1.$$

Unlike the scalar case, the sum of point values at the integers and the integral do not have to be the same. They just have to have the same inner product with $\mathbf{y}_0$.

For orthogonal ϕ, we had in the scalar case

$$\|\phi\|_2 = |\mu_0| = 1.$$

In the multiwavelet case, the corresponding normalization is

$$\|\boldsymbol{\phi}\|^2 = \sum_{k=1}^{r} \|\phi_k\|^2 = r,$$
$$\|\boldsymbol{\mu}_0\| = 1.$$

The L^2-normalization for biorthogonal multiwavelets will be given in section 7.8.1.

Example 6.6
The Chui–Lian multiwavelet CL2 has three coefficients and dilation factor 2. Its coefficients are given in appendix A. It has support on $[0, 2]$ and is symmetric/antisymmetric about $x = 1$ (fig. 6.4).

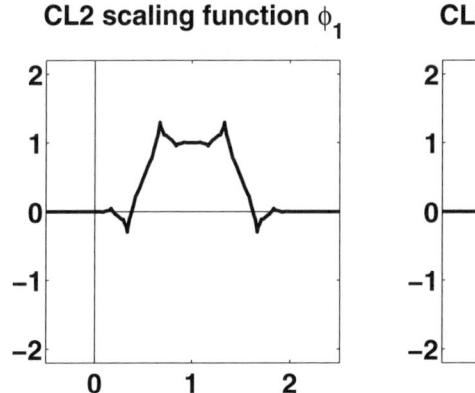

FIGURE 6.4
Two components of CL2 multiscaling function.

The matrix T has the form
$$T = \sqrt{2} \begin{pmatrix} H_0 & 0 & 0 \\ H_2 & H_1 & H_0 \\ 0 & 0 & H_2 \end{pmatrix}.$$

H_0 and H_2 both have eigenvalues $(2 - \sqrt{7})/8$ and 0, so $\phi(0) = \phi(2) = \mathbf{0}$; we can reduce the eigenvalue problem to
$$H_1 \phi(1) = \sqrt{2} \phi(1).$$
The eigenvector is
$$\phi(1) = \begin{pmatrix} 1 \\ 0 \end{pmatrix}.$$
Since the zeroth moment, normalized to $\|\boldsymbol{\mu}_0\| = 1$, is also $(1, 0)^*$, $\phi(1)$ is correctly normalized.
Then
$$\phi(1/2) = \sqrt{2} H_0 \phi(1) = \begin{pmatrix} 1/2 \\ -\sqrt{7}/4 \end{pmatrix}. \qquad \square$$

We can use the eigenvalue approach to compute point values of derivatives of ϕ (assuming they exist).

If ϕ satisfies the refinement equation

$$\phi(x) = \sqrt{m} \sum_{k=k_0}^{k_1} h_k\, \phi(mx - k),$$

then

$$(D\phi)(x) = m\sqrt{m} \sum_{k=k_0}^{k_1} h_k\, (D\phi)(mx - k).$$

With the same derivation as above, we end up with

$$D\phi = mTD\phi;$$

so $D\phi$ is an eigenvector of T to eigenvalue $1/m$. This eigenvalue must exist if ϕ is differentiable.

The nth derivative can be likewise computed from the eigenvector to eigenvalue m^{-n}.

LEMMA 6.29
The correct normalization for the nth derivative is given by

$$n! = \sum_{t=0}^{n} \binom{n}{t} \mathbf{y}_t^* \left(\sum_k D^n \phi(k) \right).$$

PROOF If ϕ has n continuous derivatives, it has approximation order at least $n+1$ (theorem 11.19), so

$$x^n = \sum_k \mathbf{c}_{nk}^* \phi(x-k)$$

$$= \sum_{t=0}^{n} \sum_k \binom{n}{t} \mathbf{y}_t^* \phi(x-k).$$

Then we just differentiate n times and set $x = 0$. ∎

REMARK 6.30 The counterpart to lemma 1.36 is also true:

$$\mathbf{y}_0^* \left(\sum_k k^n D^n \phi(k) \right) = (-1)^n n!\, \mathbf{y}_0^* \left(\int x^n D^n \phi(x)\, dx \right).$$

This may give the correct normalization, but it could reduce to $0 = 0$. The normalization given in lemma 6.29 always works. ∎

Example 6.7

For the CL2 multiscaling function, the eigenvalue problem for the derivative becomes

$$H_1 \phi'(1) = 2\sqrt{2}\phi'(1).$$

The eigenvector is $(0,1)^*$. The first two approximation vectors are

$$\mathbf{y}_0 = \begin{pmatrix} 1 \\ 0 \end{pmatrix}, \qquad \mathbf{y}_1 = \frac{1}{1+\sqrt{7}} \begin{pmatrix} 1+\sqrt{7} \\ 1 \end{pmatrix}.$$

The correct normalization for $\phi'(1)$ is

$$\phi'(1) = \begin{pmatrix} 0 \\ 1+\sqrt{7} \end{pmatrix}.$$

□

7

Practical Computation

As explained in section 6.2, the discrete multiwavelet transform (DMWT) is based on the decomposition

$$V_n = V_\ell \oplus W_\ell \oplus W_{\ell+1} \oplus \cdots \oplus W_{n-1}.$$

A function $s \in V_n$ can be expanded either as

$$s = \sum_k \mathbf{s}^*_{nk} \phi_{nk}$$

or as

$$s = \sum_k \mathbf{s}^*_{\ell k} \phi_{\ell k}(x) + \sum_{j=\ell}^{n-1} \sum_k \sum_{t=1}^{m-1} \mathbf{d}^{(t)*}_{jk} \psi^{(t)}_{jk},$$

where

$$\phi_{nk}(x) = m^{n/2} \phi(m^n x - k)$$

(and likewise for $\psi^{(t)}$, $\tilde{\phi}$, $\tilde{\psi}^{(t)}$), and

$$\mathbf{s}^*_{nk} = \langle s, \tilde{\phi}_{nk} \rangle,$$
$$\mathbf{d}^*_{nk} = \langle s, \tilde{\psi}^{(t)}_{nk} \rangle.$$

The multiscaling and multiwavelet functions are column vectors. The coefficients are row vectors.

The notation s, d originally stood for *sum* and *difference*, which is what they are for the Haar wavelet. You can also think of them as standing for the *smooth* part and the fine *detail* of s. The original function $s(x)$ is the *signal*.

The DMWT and IDMWT (inverse DMWT) convert the $\mathbf{s}_{nk}$ into $\mathbf{s}_{\ell k}$, $\mathbf{d}^{(t)}_{jk}$, $j = \ell, \ldots, n-1$, $t = 1, \ldots, m-1$, and conversely. The implementation is described in section 7.1.

The DMWT algorithm requires the initial expansion coefficients $\mathbf{s}_{nk}$. Frequently, the available data consists of equally spaced samples of s of the form $s(2^{-n}\ell)$. Converting $s(2^{-n}\ell)$ to $\mathbf{s}_{nk}$ is called *preprocessing*. After an IDMWT, converting $\mathbf{s}_{nk}$ back to $s(2^{-n}\ell)$ is called *postprocessing*. Postprocessing has to be the inverse of preprocessing if we want to achieve perfect reconstruction.

In the scalar wavelet case, the difference between function samples and expansion coefficients is not that great. Preprocessing can often be skipped without any dire consequences. In the multiwavelet case preprocessing is a necessity, not an option. We will discuss various approaches in section 7.2.

There are multiwavelets that have been specially constructed to not require a preprocessing step. We will discuss them in section 7.3.

A very important question is the handling of boundaries. The DMWT is defined for infinitely long signals. In practice, we can only handle finitely long signals. What do we do near the ends? Several approaches are covered in section 7.4.

We will then describe some alternative formulations of the DMWT algorithm, and finally some methods for computing integrals involving multiscaling or multiwavelet functions.

7.1 Discrete Multiwavelet Transform

Assume that we have a function $s \in V_n$

$$s(x) = \sum_k \mathbf{s}_{nk}^* \phi_{nk}(x),$$

represented by its coefficient sequence $\mathbf{s}_n = \{\mathbf{s}_{nk}\}$. We decompose s into its components in V_{n-1}, W_{n-1}:

$$s = P_{n-1}s + Q_{n-1}s$$
$$= \sum_j \mathbf{s}_{n-1,j}^* \phi_{n-1,j} + \sum_j \sum_{t=1}^{m-1} \mathbf{d}_{n-1,j}^{(t)*} \psi_{n-1,j}^{(t)},$$

where

$$\mathbf{s}_{n-1,j}^* = \langle s, \tilde{\phi}_{n-1,j} \rangle,$$
$$\mathbf{d}_{n-1,j}^{(t)*} = \langle s, \tilde{\psi}_{n-1,j}^{(t)} \rangle.$$

LEMMA 7.1

$$\langle \phi_{n-1,j}, \tilde{\phi}_{nk} \rangle = H_{k-mj},$$
$$\langle \phi_{n-1,j}, \tilde{\psi}_{nk}^{(t)} \rangle = G_{k-mj}^{(t)},$$
$$\langle \tilde{\phi}_{n-1,j}, \phi_{nk} \rangle = \tilde{H}_{k-mj},$$
$$\langle \tilde{\psi}_{n-1,j}^{(t)}, \phi_{nk} \rangle = \tilde{G}_{k-mj}^{(t)}.$$

PROOF We will just prove the first one:

$$\langle \phi_{n-1,j}, \tilde{\phi}_{nk} \rangle = \int m^{(n-1)/2} \phi(m^{n-1}x - j) \cdot m^{n/2} \tilde{\phi}(m^n x - k)\, dx$$

$$= \int \sum_\ell m^{n/2} H_\ell \phi(m^n x - mj - \ell) \cdot m^{n/2} \tilde{\phi}(m^n x - k)\, dx$$

$$= \sum_\ell H_\ell \delta_{mj+\ell,k} I = H_{k-mj}. \qquad \blacksquare$$

Using these formulas, we find that

$$s^*_{n-1,j} = \langle \sum_k s^*_{nk} \phi_{nk}, \tilde{\phi}_{n-1,j} \rangle$$

$$= \sum_k s^*_{nk} \langle \phi_{nk}, \tilde{\phi}_{n-1,j} \rangle$$

$$= \sum_k s^*_{nk} \tilde{H}^*_{k-mj},$$

or

$$\mathbf{s}_{n-1,j} = \sum_k \tilde{H}_{k-mj} \mathbf{s}_{nk}.$$

After similar calculations for $\mathbf{d}^{(t)}_{n-1,j}$ and for the reconstruction step, we get

ALGORITHM 7.2 Discrete Multiwavelet Transform (Direct Formulation)
The original signal is $\mathbf{s}_n$.
Decomposition:

$$\mathbf{s}_{n-1,j} = \sum_k \tilde{H}_{k-mj} \mathbf{s}_{nk},$$

$$\mathbf{d}^{(t)}_{n-1,j} = \sum_k \tilde{G}^{(t)}_{k-mj} \mathbf{s}_{nk}.$$

The decomposed signal consists of m pieces $\mathbf{s}_{n-1}$, $\mathbf{d}^{(t)}_{n-1}$, $t = 1, \ldots, m-1$.
Reconstruction:

$$\mathbf{s}_{nk} = \sum_j H^*_{k-mj} \mathbf{s}_{n-1,j} + \sum_j \sum_{t=1}^{m-1} G^{(t)*}_{k-mj} \mathbf{d}^{(t)}_{n-1,j}.$$

The formulas look identical to those in the scalar wavelet case, except for the changes due to arbitrary m, but we have to be a bit careful.

The signal $\mathbf{s}_n$ is not an infinite sequence, it is an infinite sequence of *vectors*. Each coefficient $\mathbf{s}_{nk}$ is a vector of length r, and each recursion coefficient is an $r \times r$ matrix.

We can interpret the multiwavelet algorithm in terms of convolutions and down- and upsampling as in the scalar case, but they are *block* convolutions and *block* down- and upsampling.

What we have described above is one step of the DMWT. In practice, we do this over several levels:

$$\mathbf{s}_n \to \mathbf{s}_{n-1}, \mathbf{d}^{(t)}_{n-1},$$
$$\mathbf{s}_{n-1} \to \mathbf{s}_{n-2}, \mathbf{d}^{(t)}_{n-2},$$
$$\ldots$$
$$\mathbf{s}_{\ell+1} \to \mathbf{s}_\ell, \mathbf{d}^{(t)}_\ell.$$

The IDMWT works similarly, in reverse.

The floating point count for the complete algorithm is $O(N)$, as in the scalar case.

The decomposition and reconstruction steps can be interpreted as infinite matrix–vector products. The decomposition step is

$$\begin{pmatrix} \vdots \\ \mathbf{s}_{n-1,-1} \\ \mathbf{s}_{n-1,0} \\ \mathbf{s}_{n-1,1} \\ \vdots \end{pmatrix} = \begin{pmatrix} \cdots & \cdots & & & & \\ \cdots & \tilde{H}_{m-2} & \tilde{H}_{m-1} & \tilde{H}_m & \cdots & \\ & \cdots & \tilde{H}_{-1} & \tilde{H}_0 & \tilde{H}_1 & \tilde{H}_2 & \cdots \\ & & & \cdots & \tilde{H}_{m+1} & \tilde{H}_{m+2} & \tilde{H}_{m+3} & \cdots \\ & & & & \cdots & \cdots & \end{pmatrix} \begin{pmatrix} \vdots \\ \mathbf{s}_{n,-1} \\ \mathbf{s}_{n,0} \\ \mathbf{s}_{n,1} \\ \vdots \end{pmatrix},$$

and similarly for the **d**-coefficients.

The matrix formulation becomes nicer if we interleave the **s**- and **d**-coefficients:

$$\begin{pmatrix} \vdots \\ (\mathbf{sd})_{n-1,-1} \\ (\mathbf{sd})_{n-1,0} \\ (\mathbf{sd})_{n-1,1} \\ \vdots \end{pmatrix} = \begin{pmatrix} \cdots & \cdots & \cdots & & & \\ \cdots & \tilde{L}_{-1} & \tilde{L}_0 & \tilde{L}_1 & \cdots & \\ & \cdots & \tilde{L}_{-1} & \tilde{L}_0 & \tilde{L}_1 & \cdots \\ & & \cdots & \tilde{L}_{-1} & \tilde{L}_0 & \tilde{L}_1 & \cdots \\ & & & \cdots & \cdots & \cdots & \end{pmatrix} \begin{pmatrix} \vdots \\ \mathbf{s}_{n,-1} \\ \mathbf{s}_{n,0} \\ \mathbf{s}_{n,1} \\ \vdots \end{pmatrix},$$

or simply

$$(\mathbf{sd})_{n-1} = \tilde{L}\,\mathbf{s}_n. \tag{7.1}$$

Here

$$(\mathbf{sd})_{n-1,j} = \begin{pmatrix} \mathbf{s}_{n-1,j} \\ \mathbf{d}^{(1)}_{n-1,j} \\ \vdots \\ \mathbf{d}^{(m-1)}_{n-1,j} \end{pmatrix},$$

and

$$\tilde{L}_k = \begin{pmatrix} \tilde{H}_{mk} & \tilde{H}_{mk+1} & \cdots & \tilde{H}_{mk+m-1} \\ \tilde{G}^{(1)}_{mk} & \tilde{G}^{(1)}_{mk+1} & \cdots & \tilde{G}^{(1)}_{mk+m-1} \\ \vdots & & & \vdots \\ \tilde{G}^{(m-1)}_{mk} & \tilde{G}^{(m-1)}_{mk+1} & \cdots & \tilde{G}^{(m-1)}_{mk+m-1} \end{pmatrix}.$$

$\tilde{L}$ is an infinite banded block Toeplitz matrix with blocks of size $mr \times mr$. Each block is an $m \times m$ array of $r \times r$ matrices.

The reconstruction step can be similarly written as

$$\begin{pmatrix} \vdots \\ \mathbf{s}_{n,-1} \\ \mathbf{s}_{n,0} \\ \mathbf{s}_{n,1} \\ \vdots \end{pmatrix} = \begin{pmatrix} \vdots & \vdots & & & \\ \vdots & L^*_{-1} & \vdots & & \\ \vdots & L^*_0 & L^*_{-1} & \vdots & \\ \vdots & L^*_1 & L^*_0 & L^*_{-1} & \vdots \\ & \vdots & L^*_1 & L^*_0 & \vdots \\ & & \vdots & L^*_1 & \vdots \\ & & & \vdots & \vdots \end{pmatrix} \begin{pmatrix} \vdots \\ (\mathbf{sd})_{n-1,-1} \\ (\mathbf{sd})_{n-1,0} \\ (\mathbf{sd})_{n-1,1} \\ \vdots \end{pmatrix},$$

or

$$\mathbf{s}_n = L^*(\mathbf{sd})_{n-1}. \tag{7.2}$$

The perfect reconstruction condition is expressed as

$$L^*\tilde{L} = I.$$

7.2 Pre- and Postprocessing

The DMWT algorithm requires the initial expansion coefficients $\mathbf{s}_{nk}$. Frequently, the available data consists of equally spaced samples of s. Converting the function samples to $\mathbf{s}_{nk}$ is called *preprocessing* or *prefiltering*. After an IDMWT, converting $\mathbf{s}_{nk}$ back to function values is called *postprocessing* or *postfiltering*. Postprocessing has to be the inverse of preprocessing if we want to achieve perfect reconstruction.

For scalar wavelets,

$$2^{-n/2} s(2^{-n} k) \approx s^*_{nk} = \langle s, \phi_{nk} \rangle.$$

This is not true in general for multiwavelets, with some exceptions discussed in section 7.3. Preprocessing and postprocessing steps are necessary.

A number of approaches have been proposed, and we give a short summary here.

In all the examples, we assume for simplicity that we are at level $n = 0$, and that the signal has been sampled at the points $x_{kj} = k+(j/r)$, $j = 0, \ldots, r-1$, divided into r-vectors and normalized. Thus, the data are

$$\boldsymbol{\sigma}_k = \frac{1}{\sqrt{r}} \begin{pmatrix} s(k) \\ s(k + \frac{1}{r}) \\ \vdots \\ s(k + \frac{r-1}{r}) \end{pmatrix}.$$

The reason for the normalization is the following. The expansion of the constant function 1 is

$$\sum_k \mathbf{y}_0^* \boldsymbol{\phi}(x - k) = 1.$$

For orthogonal multiwavelets, $\|\mathbf{y}_0\| = 1$. The normalization ensures that we also have $\|\boldsymbol{\sigma}_k\| = 1$.

7.2.1 Interpolating Prefilters

We try to determine multiscaling function coefficients $\mathbf{s}_k$ so that the multiscaling function series matches the function values at the points x_{kj}:

$$\sum_\ell \mathbf{s}_\ell^* \boldsymbol{\phi}(mx_{kj} - \ell) = s(x_{kj}).$$

This may or may not be possible for a given multiwavelet. This approach preserves the approximation order but not orthogonality. It is described in more detail in [153].

Example 7.1
For the DGHM multiscaling function, the only nonzero function values at the integers and half-integers are

$$\phi_1(1/2) = 4\sqrt{6}/5,$$
$$\phi_2(1/2) = \phi_2(3/2) = -3\sqrt{3}/10,$$
$$\phi_2(1) = \sqrt{3}.$$

This means that for integer k

$$s(k) = \sqrt{3} s_{k-1,2}$$
$$s(k + \frac{1}{2}) = \frac{4\sqrt{6}}{5} s_{k,1} - \frac{3\sqrt{3}}{10} s_{k,2} - \frac{3\sqrt{3}}{10} s_{k-1,2}$$
$$= \frac{4\sqrt{6}}{5} s_{k,1} - \frac{3\sqrt{3}}{10} s(k+1) - \frac{3\sqrt{3}}{10} s(k),$$

or conversely

$$s_{k,1} = \frac{5}{4\sqrt{6}} \left(\frac{3\sqrt{3}}{10} s(k) + s(k+\frac{1}{2}) + \frac{3\sqrt{3}}{10} s(k+1) \right),$$

$$s_{k,2} = s(k+1).$$

Here $s_{k,1}$ and $s_{k,2}$ are the first and second component of the 2-vector $\mathbf{s}_k$. □

[154] describes the construction of *totally interpolating* biorthogonal multiwavelets, which means that

$$\phi_j(\frac{k}{r}) = \delta_{jk},$$

and likewise for the dual and all the multiwavelet functions. Totally interpolating multiwavelets do not require a preprocessing step.

7.2.2 Quadrature-Based Prefilters

We approximate the integral defining the true coefficients by a quadrature rule. For any choice of quadrature points, we can get the weights by integrating the Lagrange interpolating polynomials at these points, as in section 2.8.3. This is explained in more detail in [94].

Example 7.2
For the DGHM multiwavelet, two cases are worked out in [94].

For one quadrature point, the weights are $\sqrt{6}/3$ for ϕ_1 and $1/\sqrt{3}$ for ϕ_2. These are just the components of the first moment $\boldsymbol{\mu}_1$. The choice of quadrature point is irrelevant.

For three quadrature points 0, 1, 2 equally spaced across the support of ϕ, we get

$$(w_{11}, w_{12}, w_{13}) = (11, 20, -3)/(14\sqrt{6}),$$
$$(w_{21}, w_{22}, w_{23}) = (-1, 30, -1)/(28\sqrt{3}).$$
□

Quadrature-based prefilters can always be found. They preserve approximation order as long as the accuracy of the quadrature rule is at least as high as the approximation order. They do not usually preserve orthogonality.

7.2.3 Hardin–Roach Prefilters

These prefilters are designed to preserve orthogonality and approximation order.

Preprocessing is assumed to be linear filtering

$$\mathbf{s}_k = \sum_{\ell} Q_{k-\ell} \boldsymbol{\sigma}_\ell,$$

or
$$\mathbf{s}(\xi) = Q(\xi)\boldsymbol{\sigma}(\xi). \tag{7.3}$$

This is an orthogonal transform if $Q(\xi)$ is paraunitary.

A *quasi-interpolating prefilter* of order p produces the correct expansion coefficients for all polynomials up to degree $p - 1$.

For the signal x^n, the point samples have the form
$$\boldsymbol{\sigma}_k = \sum_{\ell=0}^{n} \binom{n}{\ell} k^{n-\ell} \mathbf{e}_\ell,$$

where
$$\mathbf{e}_\ell = \frac{1}{\sqrt{r}} \left((\frac{0}{r})^\ell, (\frac{1}{r})^\ell, \ldots, (\frac{r-1}{r})^\ell \right)^*.$$

The true expansion coefficients have the form
$$\mathbf{s}_k = \sum_{\ell=0}^{n} \binom{n}{\ell} k^{n-\ell} \mathbf{y}_\ell,$$

where the $\mathbf{y}_\ell$ are the approximation vectors.

We can substitute this into equation (7.3) and get conditions on $Q(\xi)$.

For $p = 1$ we get
$$Q(0)\mathbf{e}_0 = \mathbf{y}_0. \tag{7.4}$$

This can always be achieved by a constant orthogonal matrix.

For $p = 2$ we get
$$Q(0)\mathbf{e}_1 - iQ'(0)\mathbf{e}_0 = \mathbf{y}_1. \tag{7.5}$$

It is proved in [76] that arbitrarily high approximation orders can be achieved if we allow a $Q(\xi)$ of sufficiently high degree.

Example 7.3
For the DGHM multiscaling function there are precisely two constant matrices Q which satisfy equation (7.4) and preserve approximation order 1:

$$\frac{1}{\sqrt{6}} \begin{pmatrix} 1+\sqrt{2} & 1-\sqrt{2} \\ -1+\sqrt{2} & 1+\sqrt{2} \end{pmatrix} \quad \text{and} \quad \frac{1}{\sqrt{6}} \begin{pmatrix} 1-\sqrt{2} & 1+\sqrt{2} \\ 1+\sqrt{2} & -1+\sqrt{2} \end{pmatrix}.$$

Approximation order 2 (equation (7.5)) can be preserved by a $Q(\xi)$ with three coefficients. There are six solutions, listed in [76]. □

An *approximation order preserving prefilter* of order p converts the point samples of any polynomial up to degree $p - 1$ into the true expansion coefficients of a polynomial of the same degree with the same leading term, but not necessarily the same polynomial. This is sufficient to achieve good results in practice.

Chapter 7: Practical Computation

For $p = 2$, equation (7.5) is replaced by

$$Q(0)\mathbf{e}_1 - iQ'(0)\mathbf{e}_0 = \mathbf{y}_1 + \alpha \mathbf{y}_0$$

for some α. Equation (7.4) remains the same.

Approximation order preserving prefilters can be shorter than quasi-interpolating prefilters.

Example 7.4
There are four orthogonal prefilters with two coefficients which preserve approximation order 2 for the DGHM multiwavelet. Details are given in [76].
◻

The Hardin–Roach approach was introduced in [76]. The details for the DGHM multiwavelet are given in that paper. Further examples are covered in [8]. Overall, only approximation order 2 has been worked out so far.

7.2.4 Other Prefilters

Other approaches to prefiltering can be found in [117], [137], [149], and [152].

We just briefly mention two other kinds of prefilters. Both of them increase the length of the input signal by a factor of r.

One method replaces the input $\{\ldots, s_0, s_1, s_2, \ldots\}$ by $\{\ldots, s_0 \mathbf{y}_0^*, s_1 \mathbf{y}_0^*, \ldots\}$.

The other method is described in [152]. We look for a function f of the form

$$f(x) = \sum_k \mathbf{a}_k^* \phi(x - k)$$

which is orthogonal to its translates, satisfies $f(0) = 1$, is low-pass, and maybe has some additional vanishing moments, so that

$$s(x) \approx \sum_k s(k) f(x - k)$$

to high accuracy. Then

$$s(x) \approx \sum_k s(k) f(x - k) = \sum_{nk} s(k) \mathbf{a}_n^* \phi(x - k - n).$$

7.3 Balanced Multiwavelets

Balanced multiwavelets are specifically constructed to not require preprocessing.

DEFINITION 7.3 *A multiwavelet is* balanced of order k *if the decomposition matrix L defined in equation (7.1) maps polynomial sequences of degree j into polynomial sequences of degree j for $j = 0, \ldots, k-1$.*

The following theorems are shown in [105].

THEOREM 7.4
A multiwavelet is balanced of order p if and only if there exist constants $r_0 = 1$, $r_1, \ldots, r_{p-1}$ so that ϕ has approximation order p with approximation vectors of the form

$$\mathbf{y}_k^* = (\rho_k(\tfrac{0}{r}), \rho_k(\tfrac{1}{r}), \ldots, \rho_k(\tfrac{r-1}{r})),$$

where

$$\rho_k(x) = \sum_{j=0}^{k} \binom{k}{j} r_{k-j} x^j.$$

THEOREM 7.5
A multiwavelet is balanced of order p if and only if the symbol factors as

$$H(\xi) = \frac{1}{m^p} C(m\xi)^p H_0(\xi) C(\xi)^{-p} \tag{7.6}$$

with

$$C(\xi) = I - \mathbf{e}_0 \mathbf{e}_0^*, \quad \mathbf{e}_0^* = \frac{1}{\sqrt{r}}(1, 1, \ldots, 1).$$

This is the two-scale similarity (TST) factorization of corollary 8.18, with $C(\xi)$ of a particular form.

Other conditions are derived in [125].

Any multiwavelet with approximation order 1 can be balanced of order 1. We replace ϕ, $\tilde{\phi}$ by

$$\phi_{\text{new}}(x) = Q\phi(x),$$
$$\tilde{\phi}_{\text{new}}(x) = Q^{-*}\tilde{\phi}(x).$$

This leaves the spaces V_0, $\tilde{V}_0$ invariant. If we choose Q so that

$$\mathbf{y}_0^* Q = \mathbf{e}_0^*,$$
$$\tilde{\mathbf{y}}_0^* Q^{-*} = \mathbf{e}_0^*,$$

the new zeroth approximation vectors will both be $\mathbf{e}_0$. Such a choice of Q is always possible. In the orthogonal case, this Q is the same Q as in the Hardin–Roach prefilter for approximation order 1.

Balancing of higher order is harder to enforce. In [13] it is shown how balancing conditions can be imposed via lifting steps. (Lifting is explained in chapter 9.)

Example 7.5
It is observed in [104] and other places that scalar orthogonal wavelets, such as the Daubechies wavelets, can be turned into balanced multiwavelets by defining

$$H_0 = \begin{pmatrix} h_0 & h_1 \\ 0 & 0 \end{pmatrix}, \quad H_1 = \begin{pmatrix} h_2 & h_3 \\ h_0 & h_1 \end{pmatrix}, \quad \ldots H_p = \begin{pmatrix} 0 & 0 \\ h_{2p-2} & h_{2p-1} \end{pmatrix}.$$

ϕ_1 is a copy of the scalar function, compressed by two, and ϕ_2 is ϕ_1 shifted right by $1/2$. □

Other examples of balanced multiwavelets are given in [13], [105], [125] to [127], and [129]. The coefficients for one of them (the BAT O1 multiwavelet from [105]) are given in appendix A.

REMARK 7.6 There are other kinds of multiwavelets which do not require preprocessing. In section 7.2.1 we already mentioned *totally interpolating multiwavelets*. Another example are the *full rank multiwavelets* from [14].

7.4 Handling Boundaries

The DMWT and IDMWT, as described so far, operate on infinite sequences of coefficients. In real life, we can only work on finite sequences. How should we handle the boundary?

Section 2.3 described the standard approach to handling boundaries in the scalar case, and three ways to look for boundary coefficients: the data extension approach, the matrix completion approach, and the boundary function approach. If you are not familiar with this material, you should go back to section 2.3 and review it first.

All of the scalar methods carry over to the multiwavelet case.

7.4.1 Data Extension Approach

The only references to the data extension approach I could find in the literature deal with symmetric extension.

We assume that there exists a point a and a matrix S with $S^2 = I$ so that
$$\phi(a + x) = S\phi(a - x), \qquad (7.7)$$
and that $\psi^{(t)}$, $\tilde{\phi}$, $\tilde{\psi}^{(t)}$ all satisfy the same relation.

For standard symmetry/antisymmetry, S is a diagonal matrix with ± 1 on the diagonal, but other choices of S are possible. For example, if $r = 2$ and
$$S = \begin{pmatrix} 0 & 1 \\ 1 & 0 \end{pmatrix},$$
then ϕ_2 is the reflection of ϕ_1 about the point a, and vice versa.

We center the functions so that $a = 0$ (whole-sample symmetry) or $a = 1/2$ (half-sample symmetry).

It is shown in [151] that in this case we can extend the signal using the same formula as in equation (7.7):
$$\mathbf{s}_{-k} = S\mathbf{s}_k \qquad \text{(whole-sample)},$$
$$\mathbf{s}_{-k} = S\mathbf{s}_{k-1} \qquad \text{(half-sample)}.$$

The finite DMWT based on symmetric extension will then preserve the symmetry across scales, as in the scalar case.

For consistency we need the additional condition
$$S\mathbf{s}_0 = \mathbf{s}_0$$
in the case of whole-sample symmetry.

In [137], the authors use an *ad hoc* symmetric extension for the DGHM multiwavelet. Since ϕ_1 has whole-sample symmetry and ϕ_2 has half-sample symmetry, the authors used corresponding extensions for the first and second components of $\mathbf{s}_k$. It seems to work in practice.

The periodic extension approach will work for multiwavelets, and will preserve orthogonality.

The other approaches (zero, constant, linear extension) given in section 2.3 appear to work in practice, based on personal numerical experimentation. I am not aware of any publications describing them.

7.4.2 Matrix Completion Approach

Personal numerical experimentation indicates that this works in the multiwavelet case. I am not aware of any publications describing it for the multiwavelet case.

7.4.3 Boundary Function Approach

As far as I know, this has only been worked out in detail for the case of the cubic Hermite multiscaling function with one particular dual. The details are given in [45], and they are quite lengthy.

Chapter 7: Practical Computation 165

There is one variation on the boundary function approach that is easy to do and has no counterpart for scalar wavelets. If we have multiwavelets with support on $[-1, 1]$, there is exactly one boundary-crossing multiscaling function at each end, and it is already orthogonal to everything inside. We can simply restrict the boundary function vector to the inside of the interval, and orthonormalize the components among themselves.

Examples for this approach are given in [72] and [75].

This would also work for scalar wavelets, of course, except there are no wavelet pairs with support in $[-1, 1]$ except the Haar wavelet. For multiwavelets, it is possible to achieve arbitrarily high approximation order, plus symmetry, on $[-1, 1]$ by taking the multiplicity high enough.

7.5 Putting It All Together

A complete DMWT for a finite one-dimensional signal goes like this:

- Do preprocessing (required except for certain types of multiwavelets).
- Decide how to handle the boundaries.
- Apply the algorithm.
- Do postprocessing.

If we start with a signal $\mathbf{s}_n$ of length N, the first decomposition step will produce m signals $\mathbf{s}_{n-1}$ and $\mathbf{d}_{n-1}^{(t)}$, $t = 1, \ldots, m-1$, each of length N/m. They can be stored in the place previously occupied by $\mathbf{s}_n$. Then we repeat the process with $\mathbf{s}_{n-1}$, and so on.

The programming is easier if we put the $\mathbf{s}_k$ at the beginning of the vector. The output from the DMWT routine after several steps is

$$\begin{pmatrix} \mathbf{s}_\ell \\ \mathbf{d}_\ell^{(1)} \\ \vdots \\ \mathbf{d}_\ell^{(m-1)} \\ \mathbf{d}_{\ell+1}^{(1)} \\ \vdots \\ \mathbf{d}_{n-1}^{(m-1)} \end{pmatrix}.$$

We can then extract the components.

Example 7.6

Figure 7.1 shows the decomposition of a signal over three levels using the DGHM multiwavelet. Only the first component of each coefficient 2-vectors is shown; a plot of the second component looks very similar. As in the scalar case, the horizontal axis has been adjusted, and the levels have been suitably scaled.

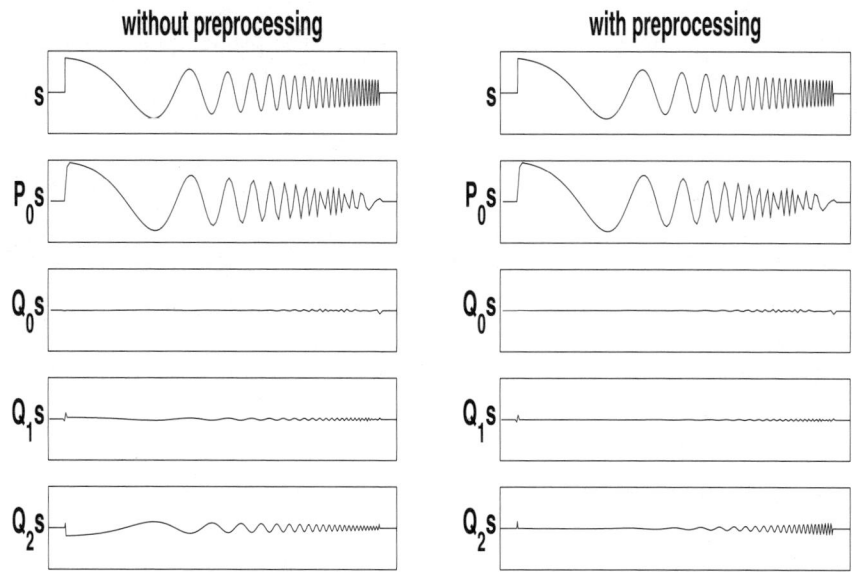

FIGURE 7.1
DWT decomposition of a signal. Left: without preprocessing. Right: with preprocessing.

The left-hand side shows the decomposition without preprocessing. On the right, the orthogonal prefilter preserving approximation order 1 was applied first. Preprocessing greatly reduces the size of the **d**-coefficients, and makes the result more comparable to figure 2.1. □

A DMWT of a two-dimensional signal (image) works the same way as for scalar wavelets: we apply the one-dimensional algorithm to the rows and to the columns. The results look similar to what is shown in chapter 2.

However, there is one extra caveat for multiwavelets. The authors of [137] determined experimentally that it is not a good idea to prefilter once and go through the entire algorithm.

What works much better is to prefilter, process, and postfilter all the rows first, and then do the same to all the columns (or vice versa).

7.6 Modulation Formulation

The modulation formulation is a way of thinking about the algorithm and verifying the perfect reconstruction conditions. It is not a way to actually implement it.

We associate with each sequence $\mathbf{a} = \{\mathbf{a}_k\}$ (finite or infinite) its symbol

$$\mathbf{a}(\xi) = \sum_k \mathbf{a}_k e^{-ik\xi}.$$

If $\mathbf{c} = \mathbf{A} * \mathbf{b}$ (convolution of a matrix and a vector sequence), then

$$\mathbf{c}(\xi) = A(\xi)\mathbf{b}(\xi).$$

Downsampling by m is represented as

$$(\downarrow m)\mathbf{c}(\xi) = \frac{1}{m} \sum_{t=0}^{m-1} \mathbf{c}\left(\frac{\xi + 2\pi t}{m}\right)$$

or

$$(\downarrow m)\mathbf{c}(m\xi) = \frac{1}{m} \sum_{t=0}^{m-1} \mathbf{c}\left(\xi + \frac{2\pi t}{m}\right).$$

Upsampling by m is represented by

$$(\uparrow m)\mathbf{c}(\xi) = \mathbf{c}(m\xi).$$

The entire DMWT algorithm in terms of the symbols is shown next.

ALGORITHM 7.7 Discrete Multiwavelet Transform (Modulation Formulation)

The original signal is $\mathbf{s}_n(\xi)$.
Decomposition:

$$\mathbf{s}_{n-1}(m\xi) = \frac{1}{\sqrt{m}} \sum_{t=0}^{m-1} \tilde{H}\left(\xi + \frac{2\pi t}{m}\right)\mathbf{s}_n\left(\xi + \frac{2\pi t}{m}\right),$$

$$\mathbf{d}_{n-1}^{(t)}(m\xi) = \frac{1}{\sqrt{m}} \sum_{t=0}^{m-1} \tilde{G}^{(t)}\left(\xi + \frac{2\pi t}{m}\right)\mathbf{s}_n\left(\xi + \frac{2\pi t}{m}\right).$$

The decomposed signal consists of m pieces $\mathbf{s}_{n-1}(m\xi)$, $\mathbf{d}^{(t)}_{n-1}(m\xi)$, $t = 1, \ldots, m-1$.

Reconstruction:

$$\frac{1}{\sqrt{m}}\mathbf{s}_n(\xi) = H(\xi)^*\mathbf{s}_{n-1}(m\xi) + \sum_{t=1}^{m-1} G^{(t)}(\xi)^* \mathbf{d}^{(t)}_{n-1}(m\xi).$$

When we add the redundant statements

$$\frac{1}{\sqrt{m}}\mathbf{s}_n(\xi + \frac{2\pi\ell}{m}) = H(\xi + \frac{2\pi\ell}{m})^*\mathbf{s}_{n-1}(m\xi) + \sum_{t=1}^{m-1} G^{(t)}(\xi + \frac{2\pi\ell}{m})^* \mathbf{d}^{(t)}_{n-1}(m\xi)$$

for $\ell = 1, \ldots, m-1$ to the reconstruction formula, we can write decomposition and reconstruction in the matrix form

$$\begin{pmatrix} s_{n-1}(m\xi) \\ d^{(1)}_{n-1}(m\xi) \\ \vdots \\ d^{(m-1)}_{n-1}(m\xi) \end{pmatrix} = \tilde{M}(\xi) \cdot \frac{1}{\sqrt{m}} \begin{pmatrix} s_n(\xi) \\ s_n(\xi + \frac{2\pi}{m}) \\ \vdots \\ s_n(\xi + \frac{(m-1)2\pi}{m}) \end{pmatrix},$$

$$\frac{1}{\sqrt{m}} \begin{pmatrix} s_n(\xi) \\ s_n(\xi + \frac{2\pi}{m}) \\ \vdots \\ s_n(\xi + \frac{(m-1)2\pi}{m}) \end{pmatrix} = M(\xi)^* \begin{pmatrix} s_{n-1}(m\xi) \\ d^{(1)}_{n-1}(m\xi) \\ \vdots \\ d^{(m-1)}_{n-1}(m\xi) \end{pmatrix}.$$

(7.8)

DEFINITION 7.8 *The matrix*

$$M(\xi) = \begin{pmatrix} H(\xi) & H(\xi + \frac{2\pi}{m}) & \cdots & H(\xi + \frac{(m-1)\pi}{m}) \\ G^{(1)}(\xi) & G^{(1)}(\xi + \frac{2\pi}{m}) & \cdots & G^{(1)}(\xi + \frac{(m-1)2\pi}{m}) \\ \vdots & & & \vdots \\ G^{(m-1)}(\xi) & G^{(m-1)}(\xi + \frac{2\pi}{m}) & \cdots & G^{(m-1)}(\xi + \frac{(m-1)2\pi}{m}) \end{pmatrix}$$

(7.9)

is called the **modulation matrix**.

The biorthogonality conditions become

$$M(\xi)^* \tilde{M}(\xi) = I. \tag{7.10}$$

The modulation matrix of an orthogonal multiwavelet is paraunitary.

REMARK 7.9 You may be wondering at this point where the factors of $1/\sqrt{m}$ come from.

Chapter 7: Practical Computation

One answer is that the symbols $H(\xi)$, $G^{(t)}(\xi)$ are defined differently from the symbols $\mathbf{s}(\xi)$, $\mathbf{d}(\xi)$: they carry a factor of $1/\sqrt{m}$ already. Together with the factor $1/\sqrt{m}$ in the decomposition formula, that makes up the factor of $1/m$ in the downsampling.

The matrix formulation makes it more clear why these factors have to be there. If $\boldsymbol{\phi}$, $\boldsymbol{\psi}^{(t)}$ form an orthogonal multiwavelet, the modulation matrix is paraunitary; it preserves 2-norms:

$$\|\mathbf{s}_n\|^2 = \|\mathbf{s}_{n-1}\|^2 + \sum_{t=1}^{m-1} \|\mathbf{d}_{n-1}^{(t)}\|^2.$$

However, at level n we also have the redundant $\mathbf{s}_n(\xi+2\ell\pi/m)$, $\ell = 1,\ldots,m-1$. The factor of $1/\sqrt{m}$ makes the two-norms of $(\mathbf{s}_n(\xi), \mathbf{s}_n(\xi + 2\pi/m), \ldots)$ and $\mathbf{s}_n(\xi)$ equal. ∎

7.7 Polyphase Formulation

There is a way to arrange the calculations in the DMWT algorithm in a form that uses block convolutions without downsampling. This is called the *polyphase implementation*. The resulting *polyphase matrix* is of great importance in the construction of multiwavelets.

DEFINITION 7.10 *The m-phases of a sequence $\mathbf{a} = \{\mathbf{a}_k\}$ are defined by*

$$\mathbf{a}_{t,k} = \mathbf{a}_{mk+t}, \quad t = 0,\ldots,m-1.$$

We split both the signal and the recursion coefficients into m-phases. Then

$$\mathbf{s}_{n-1,j} = \sum_k \tilde{H}_{k-mj} \mathbf{s}_{nk}$$

$$= \sum_k \tilde{H}_{mk-mj} \mathbf{s}_{n,mk} + \sum_k \tilde{H}_{mk+1-mj} \mathbf{s}_{n,mk+1} + \cdots$$

$$= \sum_{\ell=0}^{m-1} \sum_k \tilde{H}_{\ell,k-j} \mathbf{s}_{n\ell,k}.$$

This is now a sum of m block convolutions:

$$\mathbf{s}_{n-1} = \sum_{t=0}^{m-1} \tilde{H}(-)_t * \mathbf{s}_{nt}.$$

The other parts of the DMWT algorithm can be adapted similarly.

Decomposition begins with splitting the input into m-phases. Each phase is block filtered (convolved) with m different filters, and the results added. In the reconstruction step, we compute the phases of the result separately, and finally recombine them.

ALGORITHM 7.11 Discrete Multiwavelet Transform (Polyphase Implementation)

The original signal is $\mathbf{s}_n$.
Decomposition:
Split $\mathbf{s}_n$ into its phases $\mathbf{s}_{n,\ell}$, $\ell = 0, \ldots, m-1$.

$$s_{n-1,j} = \sum_{\ell=0}^{m-1} \sum_k \tilde{H}_{\ell,k-j} s_{n\ell,k}$$

$$d_{n-1,j}^{(t)} = \sum_{\ell=0}^{m-1} \sum_k \tilde{G}_{\ell,k-j}^{(t)} s_{n\ell,k}, \quad t = 0, \ldots, m-1.$$

Reconstruction:

$$s_{n\ell,k} = \sum_j H_{\ell,k-j}^* s_{n-1,j} + \sum_t \sum_j G_{\ell,k-j}^{(t)*} s_{n-1,j}.$$

Recombine the phases $\mathbf{s}_{n\ell}$ into $\mathbf{s}_n$.

DEFINITION 7.12 *The* polyphase symbols *of a sequence $\mathbf{a} = \{\mathbf{a}_k\}$ are given by*

$$\mathbf{a}_t(\xi) = \sum_k \mathbf{a}_{t,k} e^{-ik\xi} = \sum_k \mathbf{a}_{mk+t} e^{-ik\xi},$$

or in z-notation

$$\mathbf{a}_t(z) = \sum_k \mathbf{a}_{t,k} z^k = \sum_k \mathbf{a}_{mk+t} z^k.$$

In matrix notation, the polyphase DMWT algorithm can be written as
Decomposition:

$$\begin{pmatrix} \mathbf{s}_{n-1}(z) \\ \mathbf{d}_{n-1}^{(1)}(z) \\ \vdots \\ \mathbf{d}_{n-1}^{(m-1)}(z) \end{pmatrix} = \tilde{P}(z) \begin{pmatrix} \mathbf{s}_{n,0}(z) \\ \vdots \\ \mathbf{s}_{n,m-1}(z) \end{pmatrix}.$$

Reconstruction:

$$\begin{pmatrix} \mathbf{s}_{n,0}(z) \\ \vdots \\ \mathbf{s}_{n,m-1}(z) \end{pmatrix} = P(z)^* \begin{pmatrix} \mathbf{s}_{n-1}(z) \\ \mathbf{d}^{(1)}_{n-1}(z) \\ \vdots \\ \mathbf{d}^{(m-1)}_{n-1}(z) \end{pmatrix}.$$

DEFINITION 7.13 *The matrix*

$$P(z) = \begin{pmatrix} H_0(z) & H_1(z) & \cdots & H_{m-1}(z) \\ G^{(1)}_0(z) & G^{(1)}_1(z) & \cdots & G^{(1)}_{m-1}(z) \\ \vdots & & & \vdots \\ G^{(m-1)}_0(z) & G^{(m-1)}_1(z) & \cdots & G^{(m-1)}_{m-1}(z) \end{pmatrix}$$

is called the polyphase matrix.

The biorthogonality conditions become

$$P(z)^* \tilde{P}(z) = I. \tag{7.11}$$

REMARK 7.14 Note that the polyphase symbols of the recursion coefficients

$$H_t(\xi) = \sum_k H_{mk+t} e^{-ik\xi}$$

do *not* get a factor of $1/\sqrt{m}$ like the regular symbols. This way, the polyphase matrix of on an orthogonal multiwavelet is paraunitary, like the modulation matrix. ∎

7.8 Calculating Integrals

Multiscaling and multiwavelet functions are usually not known in closed form. Nevertheless it is possible to compute many kinds of integrals exactly or approximately.

7.8.1 Integrals with Other Refinable Functions

Assume that $\phi_1(x)$ and $\phi_2(x)$ are two refinable function vectors

$$\phi_1(x) = \sqrt{m} \sum_{k=k_0}^{k_1} H_{1k}\phi_1(mx-k),$$

$$\phi_2(x) = \sqrt{m} \sum_{\ell=\ell_0}^{\ell_1} H_{2\ell}\phi_2(mx-\ell).$$

Define

$$F(y) = \int \phi_1(x)\phi_2(x-y)^* \, dx. \tag{7.12}$$

When we substitute the refinement equations into the definition of F and sort things out, we find that

$$F(y) = \sum_{k,\ell} H_{1\ell+k} F(my-k) H_{2\ell}^*.$$

In terms of $\mathbf{F} = \text{vec}(F)$, this becomes

$$\mathbf{F}(y) = \sqrt{m} \sum_k C_k \mathbf{F}(my-k),$$

where

$$C_k = \frac{1}{\sqrt{m}} \sum_\ell \overline{H_{2\ell}} \otimes H_{1,\ell+k}.$$

Here $\overline{H_{2\ell}}$ is the element-by-element complex conjugate of $H_{2\ell}$, not the complex conjugate transpose. The Kronecker product and the *vec* operation are defined in appendix B.7.

The function vector $\mathbf{F}$ is again refinable. If $\{H_{1k}\}$ and $\{H_{2\ell}\}$ both satisfy the sum rules of order 1 with approximation vectors $\mathbf{y}_{10}$ and $\mathbf{y}_{20}$, then so does $\{C_k\}$, with approximation vector $\overline{\mathbf{y}_{20}} \otimes \mathbf{y}_{10}$.

An integral of the form

$$\int \phi_1(x)\phi_2(x-n)^* \, dx,$$

is then nothing but $\mathbf{F}(n)$, and we already know how to find point values of refinable functions.

Chapter 7: Practical Computation

For the normalization, we convert the vector $\mathbf{F}$ back into the matrix F and multiply with the zeroth approximation vectors from left and right:

$$\mathbf{y}_{10}^* \left(\sum_k F(k) \right) \mathbf{y}_{20} = \mathbf{y}_{10}^* \left(\int \sum_k \phi_1(x+k)\phi_2(x)^* \, dx \right) \mathbf{y}_{20}$$

$$= \left(\int \mathbf{y}_{10}^* \sum_k \phi_1(x+k)\phi_2(x)^* \, dx \right) \mathbf{y}_{20}$$

$$= \left(\int \phi_2(x)^* \, dx \right) \mathbf{y}_{20}$$

$$= 1.$$

As a by-product, this is also the way to compute $\|\phi\|_2$ for biorthogonal multiwavelets. We simply take $\phi_1 = \phi_2 = \phi$, normalize the point values of F at the integers correctly, and then look at $F(0)$.

Example 7.7

For the Hermit cubic multiscaling function, the nonzero recursion coefficients for $\mathbf{F}$ are

$$C_{-2} = \frac{1}{128\sqrt{2}} \begin{pmatrix} 16 & 24 & -24 & -36 \\ -4 & -4 & 6 & 6 \\ 4 & 6 & -4 & -6 \\ -1 & -1 & 1 & 1 \end{pmatrix},$$

$$C_{-1} = \frac{1}{128\sqrt{2}} \begin{pmatrix} 64 & 48 & -48 & 0 \\ -8 & 8 & 0 & -24 \\ 8 & 0 & 8 & 24 \\ 0 & 4 & -4 & -8 \end{pmatrix},$$

$$C_0 = \frac{1}{128\sqrt{2}} \begin{pmatrix} 96 & 0 & 0 & 72 \\ 0 & 24 & -12 & 0 \\ 0 & -12 & 24 & 0 \\ 2 & 0 & 0 & 18 \end{pmatrix},$$

$$C_1 = \frac{1}{128\sqrt{2}} \begin{pmatrix} 64 & -48 & 48 & 0 \\ 8 & 8 & 0 & 24 \\ -8 & 0 & 8 & -24 \\ 0 & -4 & 4 & -8 \end{pmatrix},$$

$$C_2 = \frac{1}{128\sqrt{2}} \begin{pmatrix} 16 & -24 & 24 & -36 \\ 4 & -4 & 6 & -6 \\ -4 & 6 & -4 & 6 \\ -1 & 1 & -1 & 1 \end{pmatrix}.$$

Matlab computes the eigenvector to eigenvalue 1 as

$$(0,0,0,0,-18,13/3,-13/3,1,-104,0,0,-8/3,-18,-13/3,13/3,1,0,0,0,0)^*.$$

This corresponds to

$$F(-2) = F(2) = \begin{pmatrix} 0 & 0 \\ 0 & 0 \end{pmatrix}, \quad F(-1) = \begin{pmatrix} -18 & -13/3 \\ 13/3 & 1 \end{pmatrix},$$

$$F(0) = \begin{pmatrix} -104 & 0 \\ 0 & -8/3 \end{pmatrix}, \quad F(1) = \begin{pmatrix} -18 & 13/3 \\ -13/3 & 1 \end{pmatrix},$$

so

$$\sum_k F(k) = \begin{pmatrix} -140 & 0 \\ 0 & -2/3 \end{pmatrix}.$$

Since $\mathbf{y}_0 = (1,0)^*$, we get $\mathbf{y}_0^* \sum_k F(k) \mathbf{y}_0 = -140$. We need to divide by (-140). The correctly normalized values of F are

$$F(-1) = \begin{pmatrix} 9/70 & 13/420 \\ -13/420 & -1/140 \end{pmatrix}, \quad F(0) = \begin{pmatrix} 26/35 & 0 \\ 0 & 2/105 \end{pmatrix},$$

$$F(1) = \begin{pmatrix} 9/70 & -13/420 \\ 13/420 & -1/140 \end{pmatrix}.$$

In particular,

$$\langle \phi, \phi \rangle = F(0) = \begin{pmatrix} 26/35 & 0 \\ 0 & 2/105 \end{pmatrix}.$$

This can be checked independently from equation (6.3). □

This approach can be generalized in many different ways. Everything listed in section 2.8.1 extends to the multiwavelet case. However, the details get very messy, so we will not pursue this any further.

7.8.2 Integrals with Polynomials

We already know how to calculate an integral of the form

$$\int p(x)\phi(x)\,dx$$

where p is a polynomial. This is nothing but a linear combination of the continuous moments defined in equation (6.20)

$$\mu_k = \int x^k \phi(x)\,dx, \qquad (7.13)$$

which can be calculated from equation (6.22).

7.8.3 Integrals with General Functions

Here we want to approximate integrals of the form

$$s_{0k}^* = \int s(x)\phi(x-k)^* \, dx \qquad (7.14)$$

where s is a general function. This is useful, for example, in generating the true expansion coefficients of s (see section 7.2).

The quadrature approach from section 2.8.3 carries over with little effort. This was already discussed in section 7.2.2.

The scaling function approach from section 2.8.3 also carries over in principle. In practice, I think it would be a lot more work than the quadrature approach.

7.9 Applications

This is a fairly short section. This may be a disappointment to some readers, but the fact is that most of the multiwavelet literature so far has concentrated on studying the properties of these functions. There are relatively few articles that report on actual implementations and performance.

The basic types of applications for multiwavelets are the same as for scalar wavelets. Refer to chapter 4 for an overview.

7.9.1 Signal Processing

A few studies have compared the performance of scalar wavelets and multiwavelets in image denoising and compression, including [29], [61], [63], [137], [143], and [151].

It appears that multiwavelets can do as well or better than scalar wavelets, but careful attention must be paid to preconditioning and handling of boundaries.

The authors of [137] report that filters with short support produce fewer artifacts in the reconstruction of compressed images. Multiwavelets have a definite advantage over scalar wavelets in this respect.

The use of multiwavelets for video compression is reported in the thesis of Tham [142].

7.9.2 Numerical Analysis

The main advantage of multiwavelets over scalar wavelets in numerical analysis lies in their short support, which makes boundaries much easier to handle.

For integral equations, in particular, multiwavelets with support $[0,1]$ can be used. At least some of the basis functions necessarily must be discontinuous, but for integral equations that is not a big problem.

Indeed, the first appearance of such multiwavelets was in the thesis and papers of Alpert (see [2] to [4]), before the concept of multiwavelets was invented.

Multiwavelet methods for integral equations are also discussed in [34], [114], [115], [141], and [148].

For differential equations, multiwavelets with support $[-1,1]$ can be used. Regularity and approximation order can be raised to arbitrary levels by taking the multiplicity high enough.

There is only one multiscaling function that crosses each boundary. It is already orthogonal to all the interior functions, so constructing the boundary multiscaling function is an easy matter: orthonormalize the truncated boundary-crossing multiscaling function. This automatically preserves approximation order. Finding the boundary multiwavelet function still takes a little effort.

If symmetric/antisymmetric multiwavelets are used, it is even possible to use only the antisymmetric components of the boundary function vector for problems with zero boundary conditions. Examples of suitable multiwavelets can be found in [45] and [75].

Other papers about adapting multiwavelets to the solution of differential equation include [1], [9], [10], [106], and [110].

8
Two-Scale Similarity Transforms

The two-scale similarity transform (TST) is a new, nonobvious construction for multiwavelets that has no counterpart for scalar wavelets (or rather, the concept is so trivial there that it did not need a name).

One main application is a characterization of approximation order which is useful for both theoretical and practical purposes. It leads to the counterpart of the statement "a symbol satisfies the sum rules of order p if and only if it contains a factor of $(1 + e^{-i\xi})^p$."

A second application is in the characterization of symmetry.

The material in this chapter is based mostly on the Ph.D. thesis of Strela [135], as well as the subsequent papers [121] and [136]. Some of the material in section 8.3 is previously unpublished.

8.1 Regular TSTs

Assume that ϕ is a refinable function vector, and let

$$\phi_{\text{new}}(x) = \sum_k C_k \phi(x - k)$$

for some coefficient matrices C_k. Then

$$\hat{\phi}_{\text{new}}(\xi) = C(\xi)\hat{\phi}(\xi), \tag{8.1}$$

where

$$C(\xi) = \sum_k C_k e^{-ik\xi}.$$

If $C(\xi)$ is nonsingular for all ξ, then

$$\hat{\phi}_{\text{new}}(m\xi) = C(m\xi)\hat{\phi}(m\xi) = C(m\xi)H(\xi)\hat{\phi}(\xi)$$
$$= C(m\xi)H(\xi)C(\xi)^{-1}\hat{\phi}_{\text{new}}(\xi).$$

This means that ϕ_{new} is again refinable with symbol

$$H_{\text{new}}(\xi) = C(m\xi)H(\xi)C(\xi)^{-1}.$$

177

This is a basis change which leaves all the spaces V_n invariant. We will mostly be interested in the case where there are only finitely many nonzero C_k.

If ϕ has a dual $\tilde{\phi}$, ϕ_{new} has the dual $\tilde{\phi}_{\text{new}}$ given by the symbol

$$\tilde{H}_{\text{new}}(\xi) = C(m\xi)^{-*}\tilde{H}(\xi)C(\xi).$$

This is easy to verify.

DEFINITION 8.1 *Assume $C(\xi)$ is a trigonometric matrix polynomial which is nonsingular for all ξ.*

$H_{\text{new}}(\xi)$ is a regular TST of $H(\xi)$ if

$$H_{\text{new}}(\xi) = C(m\xi)H(\xi)C(\xi)^{-1}.$$

$H_{\text{new}}(\xi)$ is a regular inverse TST (regular ITST) of $H(\xi)$ if

$$H_{\text{new}}(\xi) = C(m\xi)^{-1}H(\xi)C(\xi).$$

Since

$$H_{\text{new}}(0) = C(0)H(0)C(0)^{-1}$$

or

$$H_{\text{new}}(0) = C(0)^{-1}H(0)C(0),$$

$H_{\text{new}}(0)$ and $H(0)$ share the same eigenvalues. If $H(0)$ satisfies condition E, so does $H_{\text{new}}(0)$. However, the eigenvectors are generally different.

It is intuitively obvious that ϕ and ϕ_{new} have the same approximation order. The following theorem verifies this, and also provides formulas for computing the new approximation vectors.

THEOREM 8.2

If H_{new} is a regular TST of H, they have the same approximation order.

If $\mathbf{y}_k$ are the approximation vectors of $H(\xi)$, the corresponding approximation vectors of $H_{\text{new}}(\xi)$ are given by

$$\mathbf{y}^*_{\text{new},k} = \sum_{\ell=0}^{k} \binom{k}{\ell} i^{\ell-k}\mathbf{y}^*_\ell \left(D^{k-\ell}C^{-1}\right)(0), \quad k = 0, \ldots, p-1. \tag{8.2}$$

Likewise, if $H_{\text{new}}(\xi)$ is a regular ITST of $H(\xi)$, they have the same approximation order.

If $\mathbf{y}_k$ are the approximation vectors of $H(\xi)$, the corresponding approximation vectors of $H_{\text{new}}(\xi)$ are given by

$$\mathbf{y}^*_{\text{new},k} = \sum_{\ell=0}^{k} \binom{k}{\ell} i^{\ell-k}\mathbf{y}^*_\ell \left(D^{k-\ell}C\right)(0), \quad k = 0, \ldots, p-1. \tag{8.3}$$

PROOF We need to verify that H_{new} with the given new approximation vectors satisfies the sum rules. This is a long, tedious calculation. The details are given in [135]. ∎

Regular TSTs have applications in verifying and imposing symmetry conditions. We will do that in section 8.5.

8.2 Singular TSTs

The definition of TST and ITST makes sense in the regular case, where $C(\xi)$ is invertible for all ξ. We can also allow noninvertible $C(\xi)$ of a special type, and this is actually the more interesting application of this idea.

DEFINITION 8.3 *A TST matrix is a 2π-periodic, continuously differentiable matrix-valued function $C(\xi)$ which satisfies*

- $C(\xi)$ *is invertible for* $\xi \neq 2\pi k$, $k \in \mathbb{Z}$.
- $C(0)$ *has a simple eigenvalue 0 with left and right eigenvectors* $\mathbf{l}$ *and* $\mathbf{r}$.
- *This eigenvalue satisfies* $\lambda'(0) \neq 0$.

The last statement requires a brief explanation: as ξ varies, the eigenvalues of $C(\xi)$ vary continuously with ξ. Simple eigenvalues vary in a differentiable manner. $\lambda(\xi)$ is the eigenvalue for which $\lambda(0) = 0$. In some neighborhood of the origin, $\lambda(\xi)$ is uniquely defined and differentiable. This derivative must be nonzero at 0.

Example 8.1
The standard example is

$$C(\xi) = I - \mathbf{r}\mathbf{l}^* e^{-i\xi} = (I - \mathbf{r}\mathbf{l}^*) + \mathbf{r}\mathbf{l}^*(1 - e^{-i\xi}),$$

where $\mathbf{l}$ and $\mathbf{r}$ are normalized to $\mathbf{l}^*\mathbf{r} = 1$. The inverse is

$$C(\xi)^{-1} = (I - \mathbf{r}\mathbf{l}^*) + \frac{\mathbf{r}\mathbf{l}^*}{1 - e^{-i\xi}}, \quad \xi \neq 2\pi k.$$

Here $\mathbf{r}(\xi) = \mathbf{r}$, $\lambda(\xi) = 1 - e^{-i\xi}$, so $\lambda'(0) = i \neq 0$. □

THEOREM 8.4
Assume $C(\xi)$ is a TST matrix with eigenvectors $\mathbf{l}$, $\mathbf{r}$. Then the TST

$$H_{\text{new}}(\xi) = C(m\xi)H(\xi)C(\xi)^{-1}$$

is well-defined if and only if **r** is a right eigenvector of $H(0)$ to some eigenvalue λ_H.

If this is the case, $H_{\text{new}}(0)$ has the eigenvalue $m\lambda_H$ with left eigenvector **l**, and the other eigenvalues of $H_{\text{new}}(0)$ are the remaining eigenvalues of $H(0)$.

Remarks:
1. All the eigenvectors of $H_{\text{new}}(0)$, both left and right, will in general be different from those of $H(0)$.
2. The eigenvalue λ_H of $H(0)$ may be multiple, even degenerate. For example, if λ_H is a double eigenvalue of $H(0)$ with a single eigenvector **r**, $H_{\text{new}}(0)$ will have eigenvalues $m\lambda_H$ and λ_H, each with an eigenvector.

PROOF We only need to consider a neighborhood of $\xi = 0$, since otherwise the TST is well-defined.

The assumptions on $C(\xi)$ imply that in a neighborhood of $\xi = 0$, $C(\xi)$ has a simple eigenvalue $\lambda(\xi)$. $\lambda(\xi)$ and the right and left eigenvectors $\mathbf{r}(\xi)$, $\mathbf{l}(\xi)$ are continuously differentiable, normalized by $\mathbf{l}(\xi)^* \mathbf{r}(\xi) = 1$, and $\lambda(0) = 0$, $\mathbf{r}(0) = \mathbf{r}$, $\mathbf{l}(0) = \mathbf{l}$.

Let $V(\xi)$ be a 2π-periodic, continuously differentiable matrix whose columns are formed by $\mathbf{r}(\xi)$ and a basis for $\mathbf{l}(\xi)^\perp$. Then

$$C(\xi) = V(\xi) J(\xi) V(\xi)^{-1},$$

where

$$J(\xi) = \begin{pmatrix} \lambda(\xi) & 0 & \cdots & 0 \\ \hline 0 & & & \\ \vdots & & J_0(\xi) & \\ 0 & & & \end{pmatrix} \tag{8.4}$$

with $J_0(\xi)$ invertible for all ξ. This is a partial Jordan normal form.

Let

$$A(\xi) = V(m\xi)^{-1} H(\xi) V(\xi),$$
$$B(\xi) = J(m\xi) A(\xi) J(\xi)^{-1};$$

then

$$\begin{aligned} H_{\text{new}}(\xi) &= C(m\xi) H(\xi) C(\xi)^{-1} \\ &= V(m\xi) J(m\xi) V(m\xi)^{-1} H(\xi) V(\xi) J(\xi)^{-1} V(\xi)^{-1} \\ &= V(m\xi) B(\xi) V(\xi)^{-1}. \end{aligned}$$

$A(\xi)$ is well-defined everywhere and continuously differentiable. We split off the first row and column

$$A(\xi) = \begin{pmatrix} a(\xi) & \boldsymbol{\rho}(\xi)^* \\ \hline \boldsymbol{\gamma}(\xi) & A_0(\xi) \end{pmatrix};$$

then
$$B(\xi) = \left(\begin{array}{c|c} \lambda(m\xi)a(\xi)\lambda(\xi)^{-1} & \lambda(m\xi)\rho(\xi)^* J_0(\xi)^{-1} \\ \hline J_0(m\xi)\gamma(\xi)\lambda(\xi)^{-1} & J_0(m\xi)A_0(\xi)J_0(\xi)^{-1} \end{array}\right).$$

As $\xi \to 0$, the top left entry becomes
$$\lim_{\xi \to 0} \frac{\lambda(m\xi)}{\lambda(\xi)} a(\xi) = ma(0)$$
by l'Hôpital's rule. The remainder of the first row goes to zero.

Since $\lambda(0) = 0$, the the remainder of the first column remains bounded as $\xi \to 0$ if and only if $J_0(0)\gamma(0) = \mathbf{0}$, which is equivalent to $\gamma(0) = \mathbf{0}$. The limits can again be calculated by l'Hôpital's rule, but the values are irrelevant.

B (and therefore also H_{new}) is well-defined if and only if $\gamma(0) = \mathbf{0}$. If this is the case, $A(0)$ and $B(0)$ have the form

$$A(0) = V(0)^{-1} H(0) V(0) = \left(\begin{array}{c|c} a(0) & \rho(0)^* \\ \hline 0 & A_0(0) \end{array}\right),$$

$$B(0) = \left(\begin{array}{c|c} ma(0) & \mathbf{0}^* \\ \hline \beta(\xi) & J_0(0)A_0(0)J_0(0)^{-1} \end{array}\right).$$

This means that
$$H(0)\mathbf{r} = V(0)A(0)V(0)^{-1}\mathbf{r} = a(0)\mathbf{r},$$
so $\mathbf{r}$ is an eigenvector of $H(0)$ to eigenvalue $\lambda_H = a(0)$. Conversely, $H(0)\mathbf{r} = \lambda_H \mathbf{r}$ implies that $\gamma(0) = \mathbf{0}$, which makes the TST well-defined.

$B(0)$ has the left eigenvector $\mathbf{e}_1 = (1, 0, \ldots, 0)^*$ to eigenvalue $m\lambda_H$, which implies that
$$\mathbf{l}^* H_{\text{new}}(0) = m\lambda_H \mathbf{l}^*.$$

The remaining eigenvalues of $H_{\text{new}}(0)$ are the same as the remaining eigenvalues of $J_0(0)A_0(0)J_0^{-1}(0)$, which are the eigenvalues of $A_0(0)$, which are the remaining eigenvalues of $H(0)$. ∎

The next theorem is proved analogously.

THEOREM 8.5
Assume $C(\xi)$ is a TST matrix with eigenvectors $\mathbf{l}$, $\mathbf{r}$. Then the ITST
$$H_{\text{new}}(\xi) = C(m\xi)^{-1} H(\xi) C(\xi)$$
is well-defined if and only if $\mathbf{l}$ is a left eigenvector of $H(2\pi k/m)$, $k = 0, \ldots, m-1$, to some eigenvalue λ_k (possibly different for every k).

If this is the case, $H_{\text{new}}(2\pi k/m)$ has the eigenvalue λ_k/m with right eigenvector $\mathbf{r}$, and the other eigenvalues of $H_{\text{new}}(2\pi k/m)$ are the remaining eigenvalues of $H(2\pi k/m)$.

DEFINITION 8.6 *H_{new} is a singular TST of H if*

$$H_{\text{new}}(\xi) = \frac{1}{m} C(m\xi) H(\xi) C(\xi)^{-1} \qquad (8.5)$$

for a TST matrix $C(\xi)$ for which $C(0)$ and $H(0)$ share a common right eigenvector $\mathbf{r}$.

H_{new} is a singular ITST of H if

$$\tilde{H}_{\text{new}}(\xi) = m C(m\xi)^{-1} \tilde{H}(\xi) C(\xi)$$

for a TST matrix $C(\xi)$ for which $C(0)$ and $H(2\pi k/m)$, $k = 0, \ldots, m-1$, share a common left eigenvector $\mathbf{l}$.

Usually, we apply TSTs using the eigenvectors to eigenvalue 1 of $H(0)$. The factors of m and $1/m$ are chosen to preserve this eigenvalue. Theorems 8.4 and 8.5 show that a singular TST preserves condition E, but a singular ITST in general does not.

COROLLARY 8.7
Assume that $\boldsymbol{\phi}, \tilde{\boldsymbol{\phi}}$ are a biorthogonal pair, $\tilde{\boldsymbol{\phi}}$ has approximation order at least 1, and $C(\xi)$ is a TST matrix with right eigenvector $\mathbf{r} = \boldsymbol{\mu}_0$. Then the singular TST and ITST

$$H_{\text{new}}(\xi) = \frac{1}{m} C(m\xi) H(\xi) C(\xi)^{-1},$$
$$\tilde{H}_{\text{new}}(\xi) = m C(m\xi)^{-*} \tilde{H}(\xi) C(\xi)^{*}$$

are both well-defined, and the new symbols again satisfy the biorthogonality conditions.

PROOF The TST is well-defined by theorem 8.4.
 The sum rules in equation (6.25) for $\tilde{H}(\xi)$ for $k = 0$ imply that

$$\boldsymbol{\mu}_0^{*} \tilde{H}(0) = \boldsymbol{\mu}_0^{*},$$
$$\boldsymbol{\mu}_0^{*} \tilde{H}(2\pi k/m) = \mathbf{0}^{*}, \qquad k = 1, \ldots, m-1.$$

$\boldsymbol{\mu}_0$ is a simultaneous left eigenvector of $\tilde{H}(2\pi k/m)$ for all k, and also the left eigenvector for $C(\xi)^{*}$, on which the ITST is based. That is precisely what theorem 8.5 requires to make the ITST well-defined.
 The fact that the new multiscaling functions form another biorthogonal pair is easy to verify. ∎

An example will be given at the end of the next section.

8.3 Multiwavelet TSTs

At this point, we have only defined TSTs for the multiscaling functions. If we also have multiwavelet functions, how are they affected by a TST? This section will describe a natural way to extend TSTs to multiwavelet functions. Some of the lemmas that we prove along the way are also needed for other proofs later.

Some of the material in this section is previously unpublished.

LEMMA 8.8
If $C(\xi)$ is a TST matrix with left and right eigenvectors $\mathbf{l}$, $\mathbf{r}$, then

$$C_0(\xi) = (1 - e^{-i\xi})C(\xi)^{-1} \tag{8.6}$$

is well-defined for all ξ.

The eigenstructure of $C_0(0)$ is described by

$$C_0(0)\mathbf{r} = \frac{i}{\lambda'(0)}\mathbf{r},$$
$$C_0(0)\mathbf{v} = \mathbf{0} \quad \text{for } \mathbf{v} \in \mathbf{l}^\perp = \{\mathbf{x} : \mathbf{l}^*\mathbf{x} = 0\}.$$

This implies that for any vector $\mathbf{a}$,

$$C_0(0)\mathbf{a} = \frac{i}{\lambda'(0)}(\mathbf{l}^*\mathbf{a})\mathbf{r}.$$

PROOF We only need to check near $\xi = 0$.

Choose a basis matrix $V(\xi)$ as in the proof of theorem 8.4. From the decomposition given in equation (8.4) we get

$$C_0(\xi) = V(\xi) \begin{pmatrix} (1-e^{-i\xi})\lambda(\xi)^{-1} & 0 & \cdots & 0 \\ \hline 0 & & & \\ \vdots & & (1-e^{-i\xi})J_0(\xi)^{-1} & \\ 0 & & & \end{pmatrix} V(\xi)^{-1}.$$

As $\xi \to 0$, we get

$$C_0(0) = V(0) \begin{pmatrix} i/\lambda'(0) & 0 & \cdots & 0 \\ \hline 0 & & & \\ \vdots & & 0 & \\ 0 & & & \end{pmatrix} V(0)^{-1},$$

from which everything else follows. ∎

We note that if C has the standard form from example 8.1, then

$$C_0(\xi) = \mathbf{r}\mathbf{l}^* + (I - \mathbf{r}\mathbf{l}^*)(1 - e^{-i\xi}). \tag{8.7}$$

LEMMA 8.9
If the TST matrix $C(\xi)$ is a trigonometric polynomial with

$$\det C(\xi) = c\,(1 - e^{-i\xi}) \quad \text{for some constant } c,$$

then $C_0(\xi)$ is again a trigonometric polynomial.

PROOF $\Delta(\xi) = \det C(\xi)$ is a scalar trigonometric polynomial. It is the product of the eigenvalues, so it has a simple zero at $\xi = 0$. Therefore,

$$\Delta(\xi) = (1 - e^{-i\xi})\Delta_0(\xi)$$

with $\Delta_0(\xi) \neq 0$ for all ξ. $C(\xi)^{-1}$ can be written in explicit form as

$$C(\xi)^{-1} = \frac{1}{\Delta(\xi)} K(\xi),$$

where $K(\xi)$ is the cofactor matrix. That is, the jk-entry of $K(\xi)$ is $(-1)^{j+k}$ times the determinant of the matrix $C(\xi)$ with row k and column j removed [133].
If $\Delta(\xi) = c \cdot (1 - e^{-i\xi})$, then

$$C_0(\xi) = \frac{1}{c} K(\xi),$$

which is a trigonometric matrix polynomial. ∎

LEMMA 8.10
A trigonometric matrix polynomial

$$A(\xi) = A_0 + A_1 e^{-i\xi} + \cdots + A_n e^{-in\xi}$$

is evenly divisible by $1 - e^{-i\xi}$ if and only if

$$A(0) = \sum_k A_k = 0.$$

PROOF Long division produces

$$(1 - e^{-i\xi})^{-1} A(\xi) = A_0 + (A_0 + A_1)e^{-i\xi} + (A_0 + A_1 + A_2)e^{-2i\xi} + \cdots.$$

Chapter 8: Two-Scale Similarity Transforms

This expansion terminates if and only if $\sum A_k = 0$. ∎

COROLLARY 8.11
Let $C(\xi)$ be a TST matrix with eigenvectors $\mathbf{l}$, $\mathbf{r}$, and let $A(\xi)$ be an arbitrary trigonometric matrix polynomial. Then $A(\xi)C(\xi)^{-1}$ is well-defined if and only if
$$A(0)\mathbf{r} = \mathbf{0}.$$

PROOF We have
$$A(\xi)C(\xi)^{-1} = (1 - e^{-i\xi})^{-1} A(\xi) C_0(\xi).$$
By lemma 8.10 this is well-defined if and only if
$$A(0)C_0(0) = 0.$$
By the eigenstructure of $C_0(0)$ (lemma 8.8), this is equivalent to
$$A(0)\mathbf{r} = \mathbf{0}. \qquad \blacksquare$$

THEOREM 8.12
Assume that $\boldsymbol{\phi}$, $\boldsymbol{\psi}^{(s)}$ and $\tilde{\boldsymbol{\phi}}$, $\tilde{\boldsymbol{\psi}}^{(s)}$ are biorthogonal multiwavelets, that $\tilde{\boldsymbol{\phi}}$ has approximation order at least one, and that $C(\xi)$ is a TST matrix with right eigenvector $\mathbf{r} = \boldsymbol{\mu}_0$.
Then the extended singular TST
$$H_{\text{new}}(\xi) = \frac{1}{m} C(m\xi) H(\xi) C(\xi)^{-1},$$
$$G_{\text{new}}^{(s)}(\xi) = G^{(s)}(\xi) C(\xi)^{-1}$$
and ITST
$$\tilde{H}_{\text{new}}(\xi) = m C(m\xi)^{-*} \tilde{H}(\xi) C(\xi)^{*},$$
$$\tilde{G}_{\text{new}}^{(s)}(\xi) = \tilde{G}^{(s)}(\xi) C(\xi)^{*}$$
are well-defined and form another biorthogonal pair.

PROOF We already know that $H_{\text{new}}(\xi)$, $\tilde{H}_{\text{new}}(\xi)$ are well-defined, and biorthogonality of the new functions is easy to verify. There is no problem with $\tilde{G}_{\text{new}}^{(s)}$. It remains to show that $G_{\text{new}}^{(s)}(\xi)$ are well-defined.
Since $\tilde{\boldsymbol{\phi}}$ has approximation order at least one, the zeroth continuous moment of $\boldsymbol{\psi}^{(s)}$ vanishes for all s. By equation (6.22) this means $N_0^{(s)} \boldsymbol{\mu}_0 = G^{(s)}(0)\boldsymbol{\mu}_0 =$

$G^{(s)}(0)\mathbf{r} = \mathbf{0}$. By corollary 8.11 this is enough to ensure that $G_{\text{new}}^{(s)}(\xi)$ are well-defined. ∎

Example 8.2
For the DGHM multiwavelet, $H(0)$ has eigenvectors $\mathbf{l} = \mathbf{r} = (\sqrt{2}, 1)^*$ to eigenvalue 1. If we use the standard choice for C, we have

$$C(\xi) = I - \frac{1}{3}\begin{pmatrix} 2 & \sqrt{2} \\ \sqrt{2} & 1 \end{pmatrix} z.$$

We get new H-coefficients

$$H_{\text{new},0} = \frac{1}{360\sqrt{2}}\begin{pmatrix} 108 & 144\sqrt{2} \\ -9\sqrt{2} & -54 \end{pmatrix}, \quad H_{\text{new},1} = \frac{1}{360\sqrt{2}}\begin{pmatrix} 276 & 84\sqrt{2} \\ 57\sqrt{2} & 156 \end{pmatrix}$$

$$H_{\text{new},2} = \frac{1}{360\sqrt{2}}\begin{pmatrix} 174 & 42\sqrt{2} \\ 138\sqrt{2} & -42 \end{pmatrix}, \quad H_{\text{new},3} = \frac{1}{360\sqrt{2}}\begin{pmatrix} 18 & 12\sqrt{2} \\ 6\sqrt{2} & 18 \end{pmatrix}$$

$$H_{\text{new},4} = \frac{1}{360\sqrt{2}}\begin{pmatrix} -34 & 28\sqrt{2} \\ -17\sqrt{2} & 28 \end{pmatrix}, \quad H_{\text{new},5} = \frac{1}{360\sqrt{2}}\begin{pmatrix} 2 & -2\sqrt{2} \\ \sqrt{2} & -2 \end{pmatrix}.$$

The new G-coefficients are

$$G_{\text{new},0} = \frac{1}{60\sqrt{2}}\begin{pmatrix} -3\sqrt{2} & -18 \\ 6 & 18\sqrt{2} \end{pmatrix}, \quad G_{\text{new},1} = \frac{1}{60\sqrt{2}}\begin{pmatrix} 19\sqrt{2} & -68 \\ -38 & 8\sqrt{2} \end{pmatrix}$$

$$G_{\text{new},2} = \frac{1}{60\sqrt{2}}\begin{pmatrix} 17\sqrt{2} & -28 \\ 34 & -28\sqrt{2} \end{pmatrix}, \quad G_{\text{new},3} = \frac{1}{60\sqrt{2}}\begin{pmatrix} -\sqrt{2} & 2 \\ -2 & 2\sqrt{2} \end{pmatrix}.$$

□

Here is one more lemma we will need later.

THEOREM 8.13
Assume that ϕ is continuously differentiable with compact support and approximation order ≥ 1, and that $H(0)$ has a simple eigenvalue 1 with all other eigenvalues smaller than $1/m$ in absolute value.
If ϕ_{new} is obtained from ϕ by a singular TST

$$H_{\text{new}}(\xi) = \frac{1}{m}C(m\xi)H(\xi)C(\xi)^{-1}, \tag{8.8}$$

then

$$\hat{\phi}_{\text{new}}(\xi) = \frac{c}{i\xi}C(\xi)\hat{\phi}(\xi) \tag{8.9}$$

for some constant c.

Note the difference between equations (8.1) and (8.9).

Chapter 8: Two-Scale Similarity Transforms

It will be obvious from the proof that the constant c is not related to $C(\xi)$ or $H(\xi)$. The TST operates on symbols, not on functions. The constant c reflects the fact that the symbol only defines the multiscaling function up to an arbitrary factor; it depends on the normalizations chosen for ϕ and ϕ_{new}.

PROOF Equation (8.5) can be written as

$$C(m\xi)^{-1} H_{\text{new}}(\xi) = \frac{1}{m} H(\xi) C(\xi)^{-1}$$

or

$$C_0(m\xi) H_{\text{new}}(\xi) = \frac{1 - e^{-im\xi}}{m(1 - e^{-i\xi})} H(\xi) C_0(\xi).$$

If we take the limit as $\xi \to 0$, we get

$$C_0(0) H_{\text{new}}(0) = H(0) C_0(0).$$

We recall that

$$H_{\text{new}}(0) \hat{\phi}_{\text{new}}(0) = \hat{\phi}_{\text{new}}(0),$$
$$H(0) \hat{\phi}(0) = \hat{\phi}(0),$$

with $\hat{\phi}$, $\hat{\phi}_{\text{new}}$ unique up to a scalar multiple. Since

$$H(0) \left[C_0(0) \hat{\phi}_{\text{new}}(0) \right] = C_0(0) H_{\text{new}}(0) \hat{\phi}_{\text{new}}(0) = C_0(0) \hat{\phi}_{\text{new}}(0),$$

we must have

$$C_0 \hat{\phi}_{\text{new}}(0) = c\, \hat{\phi}(0)$$

for some constant c.

We verify

$$\prod_{k=1}^{n} H_{\text{new}}(m^{-k}\xi) = m^{-n} C(\xi) \left(\prod_{k=1}^{n} H(m^{-k}\xi) \right) C(m^{-n}\xi)^{-1}.$$

This is a telescoping product: the inner C-terms cancel.

Formally,

$$\hat{\phi}_{\text{new}}(\xi) = \lim_{n\to\infty} \left(\prod_{k=1}^{n} H_{\text{new}}(m^{-k}\xi)\right) \hat{\phi}_{\text{new}}(0)$$

$$= \lim_{n\to\infty} m^{-n} C(\xi) \prod_{k=1}^{n} H(m^{-k}\xi) C(m^{-n}\xi)^{-1} \hat{\phi}_{\text{new}}(0)$$

$$= C(\xi) \lim_{n\to\infty} \frac{m^{-n}}{1 - e^{-im^{-n}\xi}} \prod_{k=1}^{n} H(m^{-k}\xi) C_0(m^{-n}\xi) \hat{\phi}_{\text{new}}(0)$$

$$= C(\xi) \frac{1}{i\xi} \prod_{k=1}^{\infty} H(m^{-k}\xi) C_0(0) \hat{\phi}_{\text{new}}(0)$$

$$= \frac{c}{i\xi} C(\xi) \prod_{k=1}^{\infty} H(m^{-k}\xi) \hat{\phi}(0)$$

$$= \frac{c}{i\xi} C(\xi) \hat{\phi}(\xi).$$

The technical conditions given in the theorem are sufficient to ensure that these manipulations are legal. It is not clear that all of them are actually necessary. The full proof can be found in [121]. ∎

8.4 TSTs and Approximation Order

THEOREM 8.14
Assume $H(\xi)$ has approximation order $p \geq 1$ with approximation vectors $\mathbf{y}_0, \ldots, \mathbf{y}_{p-1}$. If $C(\xi)$ is a TST matrix with $\mathbf{1} = \mathbf{y}_0$, then the singular ITST

$$H_{\text{new}}(\xi) = mC(m\xi)^{-1} H(\xi) C(\xi)$$

is well-defined, and $H_{\text{new}}(\xi)$ satisfies the sum rules of order $p - 1$ with approximation vectors

$$\mathbf{y}_{\text{new},n}^* = \frac{1}{n+1} \sum_{k=0}^{n+1} \binom{n+1}{k} i^{j-n-1} \mathbf{y}_k^* (D^{n+1-k} H)(0) \qquad (8.10)$$

for $n = 0, \ldots, p - 2$, with $\mathbf{y}_{\text{new},0} \neq \mathbf{0}$.

PROOF We already know that the ITST is well-defined.

Chapter 8: Two-Scale Similarity Transforms

It remains to be verified that the new approximation vectors given in equation (8.10) satisfy the sum rules of order $p-1$ for $H_{\text{new}}(\xi)$, and that $\mathbf{y}_{\text{new},0} \neq \mathbf{0}$.
This is a lengthy calculation. The details are given in [135]. ∎

THEOREM 8.15
Assume that $H(\xi)$ has approximation order $p \geq 1$ with approximation vectors $\mathbf{y}_0, \ldots, \mathbf{y}_{p-1}$, and that $\mathbf{r}$ is a right eigenvector of $H(0)$ to eigenvalue 1.
If $C(\xi)$ is a TST matrix with right eigenvector $\mathbf{r}$, then

$$H_{\text{new}}(\xi) = \frac{1}{m} C(m\xi) H(\xi) C(\xi)^{-1}$$

is well-defined, and $H_{\text{new}}(\xi)$ satisfies the sum rules of order $p+1$.

CONJECTURE 8.16
The new approximation vectors for $H_{\text{new}}(\xi)$ in theorem 8.4 are given by

$$\mathbf{y}^*_{\text{new},k} = k\mathbf{y}^*_{k-1}[C_0(0) - i(DC_0)(0)] + \sum_{\ell=0}^{k} \binom{k}{\ell} B_{k-\ell} \mathbf{y}^*_\ell C_0(0),$$

$$k = 0, \ldots, p-1$$

$$\mathbf{y}^*_{\text{new},p} = p\mathbf{y}^*_{p-1}[C_0(0) - (DC_0)(0)] + \sum_{\ell=0}^{p-1} \binom{p}{\ell} B_{p-\ell} \mathbf{y}^*_\ell C_0(0) \qquad (8.11)$$

$$- \frac{m^p}{m^p - 1} \sum_{\ell=0}^{p-1} \binom{p}{\ell} (mi)^{\ell-p} \mathbf{y}^*_\ell (D^{p-\ell} H)(0) C_0(0).$$

Here C_0 is defined as in equation (8.6), and B_k are the Bernoulli numbers, given by

$$\frac{t}{e^t - 1} = \sum_{n=0}^{\infty} B_n \frac{t^n}{n!}.$$

Remarks about theorem 8.15 and conjecture 8.16.
The theorem is valid even in the case $p = 0$. In that case we require that $H(\xi)$ has an eigenvalue of 1 with left eigenvector $\mathbf{y}_0$ and right eigenvector $\mathbf{r}$.

TSTs in their full generality were developed in [135]. A special case which required $H(0)$ to have eigenvectors to eigenvalue 1 of a particular structure was developed independently in [120]. The two approaches were reconciled in the joint paper [121].

Among the various papers on TSTs, theorem 8.15 as well as conjecture 8.16 are proved only in [121], and only for the special case. It is the single most complicated calculation that I am aware of in all of multiwavelet theory, so it is not reproduced here.

It is shown in [121] that a general $H(\xi)$ can always be converted to one with the required eigenvector structure via a regular TST. The proof of theorem 8.15 for the special case, together with the fact that a regular TST preserves approximation order, proves that a TST raises the approximation order. It also provides formulas for the new approximation vectors which are a composition of equation (8.2) with the special case of equation (8.11). These formulas are very messy, and not particularly practical.

I have chosen to present the formulas from the special case as a conjecture for the general case.

COROLLARY 8.17

A singular TST raises the approximation order by precisely 1. A singular ITST lowers the approximation order by precisely one.

PROOF Assume that ϕ has approximation order p, but not higher. Theorem 8.15 proves that if ϕ_{new} is a singular TST of ϕ, it has approximation order at least $p+1$. If the approximation order of ϕ_{new} were higher than $p+1$, then by theorem 8.14 ϕ as the singular ITST of ϕ_{new} would have approximation order higher than p, which is a contradiction.

The other assertion is proved analogously. ∎

TSTs can be used to transfer approximation orders back and forth between a multiscaling function and its dual, similar to moving factors of $(1+e^{-i\xi})/2$ in the scalar case.

The following corollary is the main result of this chapter.

COROLLARY 8.18

If $H(\xi)$ has approximation order $p \geq 1$, it can be factored as

$$H(\xi) = H_p(\xi) = \frac{1}{m} C_p(m\xi) H_{p-1}(\xi) C_p(\xi)^{-1} = \cdots$$
$$= m^{-p} C_p(m\xi) \cdots C_1(m\xi) H_0(\xi) C_1(\xi)^{-1} \cdots C_p(\xi)^{-1}, \quad (8.12)$$

where each $C_k(\xi)$ is a TST matrix.

A natural choice is to use $C_k(\xi) = I - \mathbf{r}_k \mathbf{l}_k^* e^{-i\xi}$. In the scalar case $r = 1$, this reduces to $C_k(\xi) = 1 - e^{-i\xi}$ for all k, so

$$H(\xi) = m^{-p} \left(\frac{1-e^{-im\xi}}{1-e^{-i\xi}} \right)^p H_0(\xi).$$

This is a known formula for scalar scaling functions with dilation factor m. It is derived, for example, in [81].

Chapter 8: Two-Scale Similarity Transforms

In the case $m = 2$, this reduces further to the well-known factorization

$$H(\xi) = \left(\frac{1 + e^{-i\xi}}{2}\right)^p H_0(\xi). \tag{8.13}$$

Corollary 8.18 is the multiwavelet counterpart to equation (8.13).

Using lemma 8.8 and corollary 8.18 it is easy to show that $\det H(\xi)$ contains a factor of

$$m^{-p}\left(\frac{1 - e^{-im\xi}}{1 - e^{-i\xi}}\right)^p,$$

but in general $H(\xi)$ does not contain this scalar factor.

One can also show that

$$H_{\text{new}}(\xi) = \frac{1 - e^{-im\xi}}{m(1 - e^{-i\xi})} H(\xi)$$

produces a new refinable function vector with approximation order one higher than ϕ. However, ϕ_{new} does not have a dual in general.

Example 8.3
The Chui–Lian multiwavelet CL2 has approximation order 2. For the first step, we have $\mathbf{l} = \mathbf{r} = (1, 0)^*$. For the second step, we have $\mathbf{l} = (1 + \sqrt{7}, 1)^*$, $\mathbf{r} = (1, 0)^*$. After factoring out two approximation orders, we are left with

$$H_0(\xi) = \frac{1}{2}\begin{pmatrix} 2 & 2 \\ -\sqrt{7} & -\sqrt{7} \end{pmatrix} + \frac{1}{2(1+\sqrt{7})}\begin{pmatrix} 0 & -2 \\ 14 + 2\sqrt{7} & 2 + 3\sqrt{7} \end{pmatrix}e^{-i\xi}$$

$$+ \frac{1}{2(1+\sqrt{7})}\begin{pmatrix} -\sqrt{7} & \sqrt{7} \\ 0 & -7 - 2\sqrt{7} \end{pmatrix}e^{-2i\xi}$$

$$+ \frac{1}{2(1+\sqrt{7})^2}\begin{pmatrix} 14 + 2\sqrt{7} & \sqrt{7} \\ -28 - 16\sqrt{7} & -7 - \sqrt{7} \end{pmatrix}e^{-3i\xi}$$

$$+ \frac{1}{2(1+\sqrt{7})^2}\begin{pmatrix} -7 - \sqrt{7} & -\sqrt{7} \\ 14 + 8\sqrt{7} & 7 + \sqrt{7} \end{pmatrix}e^{-4i\xi}.$$

In the scalar case, h_0 is shorter than h. In the multiwavelet case, H_0 is often longer than H. ☐

8.5 Symmetry

DEFINITION 8.19 *A function f is symmetric about the point a if*

$$f(a + x) = f(a - x) \quad \text{for all } x.$$

f is antisymmetric *about a* if

$$f(a+x) = -f(a-x) \quad \text{for all } x.$$

On the Fourier transform side, symmetry and antisymmetry are expressed by

$$\hat{f}(\xi) = \pm e^{-2ia\xi}\hat{f}(\xi).$$

(In this formula and others, + corresponds to symmetry; − corresponds to to antisymmetry).

For a symmetric multiscaling function, each component could be symmetric or antisymmetric about a different point. In general, let

$$A(\xi) = \begin{pmatrix} \pm e^{-2ia_1\xi} & 0 & \cdots & 0 \\ 0 & \ddots & \ddots & \vdots \\ \vdots & \ddots & \ddots & 0 \\ 0 & \cdots & 0 & \pm e^{-2ia_r\xi} \end{pmatrix}. \tag{8.14}$$

If

$$\hat{\phi}(\xi) = A(\xi)\hat{\phi}(-\xi),$$

we will call ϕ "symmetric about the points a_k."

In the special case $a_j = a$ for all j,

$$A(\xi) = e^{-2ia\xi}S, \quad S = \begin{pmatrix} \pm 1 & & \\ & \ddots & \\ & & \pm 1 \end{pmatrix},$$

and the symmetry conditions are equivalent to

$$H_k = SH_{2a(m-1)-k}S.$$

LEMMA 8.20
If the symbol $H(\xi)$ satisfies

$$H(\xi) = A(m\xi)H(-\xi)A(\xi)^{-1},$$

then ϕ is symmetric about the points a_k.
If in addition

$$G^{(s)}(\xi) = B^{(s)}(m\xi)G^{(s)}(-\xi)A(\xi)^{-1},$$

where $B^{(s)}$ has the same structure as A, then $\psi^{(s)}$ is symmetric about the points $b_k^{(s)}$.

PROOF

$$\hat{\phi}(m\xi) = H(\xi)\hat{\phi}(\xi)$$
$$= A(m\xi)H(-\xi)A(\xi)^{-1}\hat{\phi}(\xi)$$
$$= A(m\xi)H(-\xi)\hat{\phi}(-\xi)$$
$$= A(m\xi)\hat{\phi}(-m\xi).$$

The second part is proved analogously. ∎

These relations can be used to test a given symbol for symmetry. In some situations they can also be used to impose desired symmetries, but there is no general algorithm.

Example 8.4
Suppose that ϕ is symmetric about the points a_k, so that

$$\hat{\phi}(\xi) = A(\xi)\hat{\phi}(-\xi).$$

We want to find a regular TST of H

$$H_{\text{new}}(\xi) = C(m\xi)H(\xi)C(\xi)^{-1}$$

so that ϕ_{new} is symmetric about some other points b_k.
Recall that $\hat{\phi}_{\text{new}}(\xi) = C(\xi)\hat{\phi}(\xi)$. Then

$$\hat{\phi}_{\text{new}}(\xi) = B(\xi)\hat{\phi}_{\text{new}}(-\xi) = B(\xi)C(-\xi)\hat{\phi}(-\xi)$$
$$= B(\xi)C(-\xi)A(\xi)^{-1}\hat{\phi}_{\text{new}}(-\xi) = C(\xi)\hat{\phi}_{\text{new}}(-\xi),$$

so we want

$$C(\xi) = B(\xi)C(-\xi)A(\xi)^{-1}, \qquad (8.15)$$

or

$$c_{k\ell}(\xi) = \pm e^{-2i(b_k - a_\ell)}c_{k\ell}(-\xi).$$

In other words, the entry $c_{k\ell}(\xi)$ must be symmetric or antisymmetric about the point $b_k - a_\ell$.

In addition, $C(\xi)$ must be a nonsingular matrix for all ξ. It may or may not be possible to fulfill these constraints in a given case. □

Example 8.5
If we make the same assumptions as in the previous example, but for a singular TST, then

$$\hat{\phi}_{\text{new}}(\xi) = \frac{c}{i\xi}C(\xi)\hat{\phi}(\xi).$$

The analogous derivation leads to

$$C(\xi) = -B(\xi)C(-\xi)A(\xi)^{-1}, \qquad (8.16)$$

which differs by a minus sign from equation (8.15). □

There is a general theorem that places limits on what is possible.

THEOREM 8.21
Assume that ϕ_{new} is a TST of ϕ, that ϕ is symmetric about the points a_k, and ϕ_{new} is symmetric about the points b_k.

If ϕ_{new} is a regular TST of ϕ, then the difference between the number of antisymmetric components in ϕ and in ϕ_{new} is odd.

If ϕ_{new} is a singular TST of ϕ, then the difference between the number of antisymmetric components in ϕ and in ϕ_{new} is odd if the multiplicity r is even, and even if r is odd.

PROOF We know that

$$\det C(\xi) = (1 - e^{-i\xi})\Delta_0(\xi) \qquad (8.17)$$

with $\Delta_0(\xi) \neq 0$ (see proof of 7.24). For the regular TST case, equation (8.15) gives

$$\det C(\xi) = \det B(\xi) \det A(\xi)^{-1} \det C(-\xi)$$
$$= e^{-2i\sum(b_k-a_k)\xi}(-1)^N \det C(-\xi),$$

where N is the difference between the number of antisymmetric components.
Using equation (8.17), this leads to

$$\Delta_0(\xi) = (-1)^{N+1} e^{-(2\sum(b_k-a_k)-1)i\xi} \Delta_0(-\xi),$$

which says

$$\Delta_0(0) = (-1)^{N+1}\Delta_0(0) \neq 0.$$

This is only possible if $N+1$ is even.

The second part is proved the same way, except that the minus sign in equation (8.16) adds an additional factor of $(-1)^r$; thus, the conclusion is that $N+r+1$ must be even. ∎

Example 8.6
The multiscaling function components of the DGHM multiwavelet are symmetric about the points $1/2$ and 1. We want to apply a TST step which makes

Chapter 8: Two-Scale Similarity Transforms

the new multiscaling functions symmetric about the point 1. By theorem 8.21, one of them will be symmetric, and one will be antisymmetric.

A TST matrix which satisfies the symmetry conditions and has the correct eigenvector **r** is

$$C(\xi) = \begin{pmatrix} 1 + e^{-i\xi} & -2\sqrt{2} \\ 1 - e^{-i\xi} & 0 \end{pmatrix}.$$

This was found by trial and error.

The new symbol is

$$H_{\text{new}}(\xi) = \frac{1}{40} \begin{pmatrix} -7 + 10e^{-i\xi} - 7e^{-2i\xi} & 15 - 15e^{-2i\xi} \\ -4 + 4e^{-2i\xi} & 10 + 20e^{-i\xi} + 10e^{-2i\xi} \end{pmatrix},$$

which is actually shorter than H. ☐

REMARK 8.22 More general types of symmetry are possible. In [151] the authors only require $S^2 = I$ instead of $S = \text{diag}(\pm 1, \pm 1, \dots)$,

For example, if

$$A = e^{-2ia\xi} I, \qquad S = \begin{pmatrix} 0 & 1 \\ 1 & 0 \end{pmatrix},$$

then ϕ_2 is the reflection of ϕ_1 about $x = a$, and vice versa.

The constructions in this chapter carry over to the more general case. ∎

9

Factorizations of Polyphase Matrices

There is a distinct difference between the modulation matrix and the polyphase matrix: the modulation matrix has a particular structure. All the information is already contained in the first column; the other columns are simply the first column with shifted argument.

If we multiply the modulation matrices of different multiwavelets together, the result does not have any particular significance. Conversely, if we want to multiply a modulation matrix by some factor to create another modulation matrix, that factor has to have a particular structure.

A polyphase matrix, on the other hand, is unstructured. If P_1, $\tilde{P}_1$ and P_2, $\tilde{P}_2$ both satisfy
$$P_k(\xi)^* \tilde{P}_k(\xi) = I,$$
then so do $P = P_1 P_2$ and $\tilde{P} = \tilde{P}_1 \tilde{P}_2$.

This makes it possible to create new multiwavelets from existing ones by multiplying the polyphase matrix by some appropriate factor, and it opens the possibility of factoring a given polyphase matrix into elementary steps. Two such factorizations are described in this chapter: one based on projection factors, and one based on lifting steps.

For simplicity, we switch to the z-notation at this point, where
$$z = e^{-i\xi}.$$
This lets us work with polynomials rather than trigonometric polynomials.

9.1 Projection Factors

9.1.1 Orthogonal Case

We begin with the orthogonal case. At the algebraic level, the polyphase matrix of an orthogonal multiwavelet is just a paraunitary matrix $P(z)$. If we multiply it by any other paraunitary matrix $F(z)$, then $P(z)F(z)$ is again paraunitary and can be interpreted as the polyphase matrix of a new orthogonal multiwavelet.

This approach will always lead to new filterbanks. Whether the new polyphase matrix actually defines a new multiscaling function which exists as a function or generates an MRA needs to be checked after the fact. I am not aware of any conditions on $F(z)$ that will guarantee this automatically.

We could also consider multiplication on the left: $F(z)P(z)$. That works equally well, but multiplication on the right has the advantage that it can be applied when only the first row of $P(z)$ is known (i.e., we know ϕ, but not $\psi^{(s)}$).

The simplest choice for a factor is $F(z) = Q$, a constant unitary matrix, but that will not get us very far. We consider the next simplest case, where $F(z)$ is linear.

If we assume $F(z) = A + Bz$, then $Q = F(1) = A + B$ must be unitary. We pull out the constant factor Q and assume $A + B = I$, so

$$F(z) = (I - B) + Bz.$$

We want $F(z)$ to be paraunitary. This requires $B^* = B$ and

$$I = F(z)F(z)^* = [(I - B) + Bz][(I - B) + Bz^{-1}]$$
$$= (I - 2B + 2B^2) + (B - B^2)(z + z^{-1}).$$

This is satisfied if and only if $B^2 = B$, which means B must be an orthogonal projection onto some subspace.

DEFINITION 9.1 *An* orthogonal projection factor *of rank k is a linear paraunitary matrix of the form*

$$F(z) = (I - UU^*) + UU^*z,$$

where the columns of U form an orthonormal basis of some k-dimensional subspace.

We list some properties that are easy to verify:

- $\det F(z) = z^k$.

- If $k \geq 2$, we can split U into two matrices $U_1 = (\mathbf{u}_1, \ldots, \mathbf{u}_\ell)$, $U_2 = (\mathbf{u}_{\ell+1}, \ldots, \mathbf{u}_k)$, and

$$[(I - UU^*) + UU^*z] = [(I - U_1U_1^*) + U_1U_1^*z][(I - U_2U_2^*) + U_2U_2^*z].$$

  In particular, any orthogonal projection factor of rank k can be written as a product of k projection factors of rank 1.

- $F(z)$ depends only on the subspace spanned by the columns of U, not on the choice of basis.

THEOREM 9.2

Assume that $P(z)$ is the polyphase matrix of an orthogonal multiwavelet, normalized so that the subscripts start at 0:

$$P(z) = P_0 + P_1 z + \cdots + P_n z^n,$$

with $P_0 \neq 0$, $P_n \neq 0$.

Then $P(z)$ can be factored in the form

$$P(z) = Q F_1(z) \cdots F_n(z),$$

where Q is a constant unitary matrix, and each $F_j(z)$ is a projection factor. The number of factors n equals the degree of $P(z)$.

REMARK 9.3 This theorem has been rediscovered multiple times, and still seems to be relatively unknown. It appeared in [146]; later, it was published again in [95], [124], [132], [144], and probably other places.

Vaidyanathan [146] uses only rank 1 projection factors. In that case, the number of projection factors equals the degree of $\det P(z)$, which is also called the *McMillan degree*. ∎

PROOF We denote the range of a matrix A by $R(A)$, its nullspace by $N(A)$. A well-known identity from linear algebra states that

$$R(A^*) = N(A)^\perp.$$

We will use that frequently in this section.

For $n > 0$, we consider

$$(P_0 + P_1 z + \cdots + P_n z^n)((I - UU^*) + UU^* z^{-1}).$$

We want to choose U so that the product is a matrix polynomial of degree $n - 1$. This requires

$$\begin{aligned} P_0 UU^* &= 0, \\ P_n(I - UU^*) &= 0. \end{aligned} \quad (9.1)$$

The fact that $P(z)$ is paraunitary implies $P_0 P_n^* = 0$, so

$$R(P_n^*) \subset N(P_0).$$

Choose any matrix U with orthonormal columns so that

$$R(P_n^*) \subset R(U) \subset N(P_0).$$

Then $P_0 U = 0$ and $UU^* P_n^* = P_n^*$, so equation (9.1) is satisfied. We choose

$$F_n(z) = ((I - UU^*) + UU^* z)$$

and repeat the process, until we get to

$$P(z)F_n^*(z) \cdots F_1^*(z) = Q.$$ ∎

Theorem 9.2 was already used in chapter 3 for alternative constructions of Daubechies wavelets and coiflets. We will see another use for it in chapter 10, where it will be used for the orthogonal completion problem (given ϕ, find the multiwavelet functions).

REMARK 9.4 Approximation order 1 (or higher) can be enforced by a proper choice of Q.

Refer to theorem 6.9 (iv): approximation order 1 means that the approximation vector $\mathbf{y}_0$ is a left eigenvector of $H_k(0)$ (the kth polyphase symbol) to eigenvalue $1/\sqrt{m}$, for all k. Since the projection factors all satisfy $F(0) = I$, it is up to Q to satisfy the first approximation order condition. All $r \times r$ blocks in the top row of Q must have a common left eigenvector to eigenvalue $1/\sqrt{m}$.

In the scalar case, this reduces to

$$Q = \frac{1}{\sqrt{2}} \begin{pmatrix} 1 & 1 \\ \pm 1 & \mp 1 \end{pmatrix}.$$

∎

9.1.2 Biorthogonal Case

In the biorthogonal case, we again look for factors of the form $F(z) = A + Bz$ with a dual of the form $\tilde{F}(z) = \tilde{A} + \tilde{B}z$. With the same normalization $A + B = I$, we find that we still need $B^2 = B$, but not $B^* = B$. B is a nonorthogonal projection.

DEFINITION 9.5 *A biorthogonal projection factor is a linear matrix polynomial of the form*

$$F(z) = (I - UV^*) + UV^*z, \tag{9.2}$$

where the columns of U, V form biorthogonal bases of two k-dimensional subspaces. This means that

$$V^*U = U^*V = I.$$

The dual of $F(z)$ is given by

$$\tilde{F}(z) = (I - VU^*) + VU^*z.$$

We get similar properties as in the orthogonal case:

- $\det F(z) = cz^k$ for some nonzero constant c.

- If $k \geq 2$, we can split U into two matrices $U_1 = (\mathbf{u}_1, \ldots, \mathbf{u}_\ell)$, $U_2 = (\mathbf{u}_{\ell+1}, \ldots, \mathbf{u}_k)$, and similarly for V, and
$$[(I - UV^*) + UV^*z] = [(I - U_1V_1^*) + U_1V_1^*z][(I - U_2V_2^*) + U_2V_2^*z].$$
  In particular, any biorthogonal projection factor of rank k can be written as a product of k projection factors of rank 1.

- $F(z)$ depends only on the subspaces spanned by the columns of U, V, not on the individual basis vectors.

Can we use the same approach as in theorem 9.2 to factor biorthogonal polyphase matrices? Unfortunately, we cannot.

THEOREM 9.6
Assume that P, $\tilde{P}$ are the polyphase matrices of a biorthogonal multiwavelet pair, normalized so that the subscripts of P start at 0:
$$P(z) = P_0 + P_1 z + \cdots + P_n z^n,$$
$$\tilde{P}(z) = \tilde{P}_k z^k + \cdots + \tilde{P}_\ell z^\ell,$$
with P_0, P_n, $\tilde{P}_k$, $\tilde{P}_\ell$ nonzero.

There exists a biorthogonal projection factor $F(z)$ so that multiplication by $F(z)^{-1}$ and $\tilde{F}(z)^{-1}$, respectively, lowers the degree of both $P(z)$ and $\tilde{P}(z)$, if and only if
$$\begin{aligned} R(\tilde{P}_\ell^*) &\subset R(U) \subset N(P_0), \\ R(\tilde{P}_k^*) &\subset R(V)^\perp \subset N(P_n). \end{aligned} \tag{9.3}$$

The proof is analogous to that of theorem 9.2.

The condition in equation (9.3) is rather specialized. A sufficient condition for the existence of suitable U, V is
$$R(P_0) \oplus R(P_n) = \mathbb{C}^{mr}.$$

In that case we can use $R(U) = R(P_0)^\perp$, $R(V) = R(P_n)$.

We cannot expect to decompose a biorthogonal polyphase matrix entirely into projection factors. A more useful approach is based on the following lemma.

LEMMA 9.7
Let $P(z)$, $\tilde{P}(z)$ satisfy the same conditions as in theorem 9.6.

If $\ell \geq 1$, we can find an orthogonal projection factor $F(z)$ so that $P(z)F(z)^{-1}$ has the same range of subscripts as before, but the range of subscripts of $\tilde{P}(z)F(z)^{-1}$ is $k-1, \ldots, \ell-1$.

PROOF Expand the biorthogonality relation $P(z)\tilde{P}(z)^* = I$. The term with lowest exponent is $P_0 \tilde{P}_\ell^* z^{-\ell}$. This implies that $\ell \geq 0$; if $\ell > 0$, then $P_0 \tilde{P}_\ell^* = 0$.

With the same type of arguments as in the proof of theorem 9.2 we verify that if we choose U to satisfy

$$R(\tilde{P}_\ell^*) \subset R(U) \subset N(P_0), \tag{9.4}$$

then the resulting $F(z)$ has the desired effect. ∎

We could equally well use a biorthogonal projection factor in the preceding theorem. The choice of V is irrelevant.

Lemma 9.7 says that as long as $\ell > 0$, we can apply an orthogonal projection factor and shift $\tilde{P}$ to the left. What happens once we get to $\ell = 0$?

At that point it is easier to consider $P(z)^{-1} = \tilde{P}(z)^*$ rather than $\tilde{P}(z)$. Since $P_0 \tilde{P}_0^* = I$, we can factor out the constant terms and assume

$$P(z) = I + P_1 z + \cdots + P_n z^n,$$
$$P(z)^{-1} = I + \tilde{P}_1^* z + \cdots + \tilde{P}_k^* z^k.$$

The determinants of both $P(z)$, $P(z)^{-1}$ are monomials in z whose product is 1, so they are constants. By letting $z \to 0$, we see that the constants have to be 1. Both $P(z)$ and $P(z)^{-1}$ are *unimodular* and *co-monic*.

In linear algebra terms, "unimodular" means "having determinant 1." "Co-monic" means that the coefficient of z^0 is I. (Monic means that the highest power has coefficient I.)

DEFINITION 9.8 *A unimodular, co-monic pair of matrix polynomials $P(z)$, $P(z)^{-1}$ is called an* atom.

This name is taken from [95]. In [123], atoms are called *pseudo-identity matrix pairs*.

We have proved the following theorem.

THEOREM 9.9
Let $P(z)$, $\tilde{P}(z)$ satisfy the same conditions as in theorem 9.6.
  Then $P(z)$ can be factored in the form

$$P(z) = CA(z)F_1(z) \cdots F_\ell(z),$$

Chapter 9: Factorizations of Polyphase Matrices

where each $F_j(z)$ is an orthogonal projection factor, $A(z)$ is part of an atom, and C is a constant matrix.

Example 9.1
The Cohen(2,2) scalar wavelet has the polyphase matrices

$$P(z) = \frac{1}{4\sqrt{2}}\left[\begin{pmatrix} 0 & 2 \\ 0 & -1 \end{pmatrix} z^{-1} + \begin{pmatrix} 4 & 2 \\ -2 & 6 \end{pmatrix} + \begin{pmatrix} 0 & 0 \\ -2 & -1 \end{pmatrix} z\right],$$

$$\tilde{P}(z) = \frac{1}{4\sqrt{2}}\left[\begin{pmatrix} -1 & 2 \\ 0 & 0 \end{pmatrix} z^{-1} + \begin{pmatrix} 6 & 2 \\ -2 & 4 \end{pmatrix} + \begin{pmatrix} -1 & 0 \\ -2 & 0 \end{pmatrix} z\right].$$

We apply a shift so that both of them start with a z^0-term. We can then factor out two orthogonal projection factors, based on the vectors

$$\mathbf{u}_1 = \begin{pmatrix} 1 \\ 0 \end{pmatrix}, \qquad \mathbf{u}_2 = \frac{1}{\sqrt{5}}\begin{pmatrix} 1 \\ 2 \end{pmatrix}.$$

At this point, P has indices 0, 1, 2, and $\tilde{P}$ has indices $-2, -1, 0$: we have reached an atom.

After we pull out the constant matrix C, what remains is

$$A(z) = I + \frac{1}{80}\begin{pmatrix} 14 & -4 \\ 9 & -14 \end{pmatrix} z + \frac{1}{80}\begin{pmatrix} 0 & 0 \\ -7 & 2 \end{pmatrix} z^2,$$

$$\tilde{A}(z)^* = I + \frac{1}{80}\begin{pmatrix} -14 & 4 \\ -9 & 14 \end{pmatrix} z + \frac{1}{80}\begin{pmatrix} 2 & 0 \\ 7 & 0 \end{pmatrix} z^2.$$
□

Theorem 9.9 says that any biorthogonal multiwavelet can be constructed from an atom and several orthogonal projection factors. A complete description of the structure of atoms, and a method for constructing all possible atoms of given type from scratch, are given in [96]. Together with the projection factors, which are easy to construct, we can build biorthogonal multiwavelets of any size.

9.2 Lifting Steps

In the scalar case with dilation factor $m = 2$, a lifting step was defined as

$$P_{\text{new}}(z) = \begin{pmatrix} 1 & a(z) \\ 0 & 1 \end{pmatrix} \begin{pmatrix} H_0(z) & H_1(z) \\ G_0(z) & G_1(z) \end{pmatrix},$$

a dual lifting step as

$$P_{\text{new}}(z) = \begin{pmatrix} 1 & 0 \\ b(z) & 1 \end{pmatrix} \begin{pmatrix} H_0(z) & H_1(z) \\ G_0(z) & G_1(z) \end{pmatrix}.$$

The lifting matrices are upper or lower unit triangular, and preserve either the multiscaling function or dual multiscaling function while changing everything else.

There are various ways to generalize this to the multiwavelet setting. The most general approach defines a lifting step as

$$P_{\text{new}}(z) = \begin{pmatrix} S(z) & L(z) \\ 0 & T(z) \end{pmatrix} P(z),$$

where $S(z)$ and $T(z)$ are invertible matrices. $S(z)$ is of size $r \times r$ and acts on the multiscaling function alone; $T(z)$ is of size $(m-1)r \times (m-1)r$ and jointly acts on all the multiwavelet functions.

Such general lifting matrices can always be factored into simpler pieces:

$$\begin{pmatrix} S & 0 \\ L & T \end{pmatrix} = \begin{pmatrix} S & 0 \\ 0 & T \end{pmatrix} \begin{pmatrix} I & 0 \\ T^{-1}L & I \end{pmatrix},$$

$$\begin{pmatrix} S & L \\ 0 & T \end{pmatrix} = \begin{pmatrix} S & 0 \\ 0 & T \end{pmatrix} \begin{pmatrix} I & S^{-1}L \\ 0 & I \end{pmatrix}.$$

The relations

$$\begin{pmatrix} I & 0 \\ L & I \end{pmatrix} \begin{pmatrix} S & 0 \\ 0 & T \end{pmatrix} = \begin{pmatrix} S & 0 \\ 0 & T \end{pmatrix} \begin{pmatrix} I & 0 \\ T^{-1}LS & I \end{pmatrix},$$

$$\begin{pmatrix} I & L^* \\ 0 & I \end{pmatrix} \begin{pmatrix} S & 0 \\ 0 & T \end{pmatrix} = \begin{pmatrix} S & 0 \\ 0 & T \end{pmatrix} \begin{pmatrix} I & S^{-1}L^*T \\ 0 & I \end{pmatrix}$$

allow us to collect all the diagonal terms at the beginning or end of a sequence of lifting steps.

THEOREM 9.10
The polyphase matrix of any biorthogonal multiwavelet can be factored in the form

$$P(z) = \begin{pmatrix} S(z) & 0 \\ 0 & T(z) \end{pmatrix} \prod_{k=0}^{N} \begin{pmatrix} I & 0 \\ L_{2k-1}(z) & I \end{pmatrix} \begin{pmatrix} I & L_{2k}(z)^* \\ 0 & I \end{pmatrix},$$

where $S(z)$, $T(z)$ are invertible matrices. $S(z)$ and $T(z)$ can be further factored into a product of a diagonal matrix with monomial entries, and several upper and lower unit triangular matrices

$$S(z) = D_S(z) \prod_k L_{S,k}(z) U_{S,k}(z),$$

$$T(z) = D_T(z) \prod_k L_{T,k}(z) U_{T,k}(z).$$

This is proved in [54] and [67].

Chapter 9: Factorizations of Polyphase Matrices

In this chapter, we will use a narrower definition of a lifting step.

DEFINITION 9.11 *A multiwavelet lifting step is defined as*

$$P_{\text{new}}(z) = \begin{pmatrix} I & L^{(1)}(z) & \cdots & \cdots & L^{(m-1)}(z) \\ 0 & I & 0 & \cdots & 0 \\ \vdots & 0 & \ddots & \ddots & \vdots \\ \vdots & \vdots & \ddots & I & 0 \\ 0 & 0 & \cdots & 0 & I \end{pmatrix} P(z).$$

A multiwavelet dual lifting step is defined as

$$P_{\text{new}}(z) = \begin{pmatrix} I & 0 & \cdots & \cdots & 0 \\ L^{(1)}(z) & I & 0 & \cdots & 0 \\ \vdots & 0 & \ddots & \ddots & \vdots \\ \vdots & \vdots & \ddots & I & 0 \\ L^{(m-1)}(z) & 0 & \cdots & 0 & I \end{pmatrix} P(z).$$

There is no consensus on this in the literature, however. The example in [54] uses a dual lifting step of the form

$$P_{\text{new}}(z) = \begin{pmatrix} I & 0 \\ L(z) & T(z) \end{pmatrix} P(z)$$

with unit upper triangular $T(z)$.

Definition 9.11 achieves the mixing of multiscaling and multiwavelet functions, which is the most important aspect of lifting, and it has the advantage that the inverse is easy to write in closed form.

Written in detail, the effect of a lifting step on the polyphase symbols is

$$H_{\text{new},k}(z) = H_k(z) + \sum_{j=1}^{m-1} L^{(j)}(z) G_k^{(j)}(z),$$

$$G_{\text{new},k}^{(j)}(z) = G_k^{(j)}(z)$$

$$\tilde{H}_{\text{new},k}(z) = \tilde{H}_k(z)$$

$$\tilde{G}_{\text{new},k}^{(j)}(z) = \tilde{G}_k^{(j)}(z) - \sum_{j=1}^{m-1} L^{(j)}(z)^* \tilde{G}_k^{(j)}(z),$$

or in terms of the symbols

$$H_{\text{new}}(z) = H(z) + \sum_{j=1}^{m-1} L^{(j)}(z^m) G^{(j)}(z),$$

$$G^{(j)}_{\text{new}}(z) = G^{(j)}(z),$$

$$\tilde{H}_{\text{new}}(z) = \tilde{H}(z),$$

$$\tilde{G}^{(j)}_{\text{new}}(z) = \tilde{G}^{(j)}(z) - \sum_{j=1}^{m-1} L^{(j)}(z^m) \tilde{G}^{(j)}(z).$$

REMARK 9.12   1. Lifting does not preserve orthogonality. If ϕ is orthogonal, the new pair ϕ, $\tilde{\phi}$ will never be orthogonal.

2. Lifting mixes ϕ, $\psi^{(s)}$, so it requires the full multiwavelet. Lifting cannot be done on ϕ alone.

3. Lifting can be used to raise the approximation order. More precisely, lifting can raise the approximation order of ϕ, and a dual lifting step can raise the approximation order of $\tilde{\phi}$. This will be described in more detail next. ∎

Papers on multiwavelet lifting include [12], [54], and [67].

A lifting procedure which imposes symmetry conditions is described in [145]. A lifting procedure which imposes balancing conditions is described in [13].

9.3 Raising Approximation Order by Lifting

The material in this subsection is based on [98].

Our goal is to raise the approximation order of a given multiwavelet pair ϕ, $\tilde{\phi}$ by lifting. This requires two steps, one for ϕ and one for $\tilde{\phi}$.

We describe the dual step, since it is easier. The primary lifting step is done by reversing the roles of ϕ, $\tilde{\phi}$, and applying another dual lifting step.

THEOREM 9.13
Assume ϕ, $\tilde{\phi}$ form a biorthogonal multiwavelet pair, and that $H(0)$ satisfies condition E. For any given $\tilde{p} \geq 1$ it is possible to use a single dual lifting step to raise the approximation order of $\tilde{\phi}$ to $\tilde{p}$.

The lifting factors $L^{(k)}(z)$ can always be chosen to have degree at most $\tilde{p}-1$, but in most cases degree $\lceil \tilde{p}/r \rceil - 1$ will suffice.

The symbol $\lceil x \rceil$ denotes the smallest integer greater than or equal to x.

Chapter 9: Factorizations of Polyphase Matrices

PROOF This is a sketch of the proof. The full proof, along with an efficient implementation, can be found in [98].

In this section we use the ξ notation, since we need to take derivatives.

By theorem 6.27 (vi), we need to make sure that the new multiwavelet functions have $\tilde{p}$ vanishing continuous moments:

$$\boldsymbol{\nu}^{(t)}_{\text{new},k} = \mathbf{0}, \quad k = 0, \ldots, \tilde{p} - 1, \quad t = 1, \ldots, m - 1.$$

By equation (6.22), these moments are expressed as

$$\boldsymbol{\nu}^{(t)}_{\text{new},k} = m^{-k} \sum_{s=0}^{k} \binom{k}{s} N^{(t)}_{\text{new},s} \boldsymbol{\mu}_{k-s}. \tag{9.5}$$

If

$$L^{(t)}(\xi) = \sum_{k} L^{(t)}_{k} e^{-ik\xi},$$

the discrete moments of $L^{(t)}$ are

$$\Lambda^{(t)}_{\ell} = \sum_{j} j^{\ell} L^{(t)}_{j}.$$

The new multiscaling function symbols are

$$G^{(t)}_{\text{new}}(\xi) = G^{(t)}(\xi) + L^{(t)}(m\xi) H(\xi).$$

This leads to

$$N^{(t)}_{\text{new},s} = N^{(t)}_{s} + \sum_{\ell=0}^{s} \binom{s}{\ell} m^{\ell} \Lambda^{(t)}_{\ell} M_{s-\ell}.$$

Putting all these things together, we get

$$\mathbf{0} = \sum_{s=0}^{k} \binom{k}{s} \left[N^{(t)}_{s} + \sum_{\ell=0}^{s} \binom{s}{\ell} m^{\ell} \left(\sum_{j} j^{\ell} L^{(t)}_{j} \right) M_{s-\ell} \right] \boldsymbol{\mu}_{k-s}$$

$$= \sum_{s=0}^{k} \binom{k}{s} N^{(t)}_{s} \boldsymbol{\mu}_{k-s} + \sum_{s=0}^{k} \sum_{\ell=0}^{s} \sum_{j} \binom{k}{s} \binom{s}{\ell} m^{\ell} j^{\ell} L^{(t)}_{j} M_{s-\ell} \boldsymbol{\mu}_{k-s}.$$

After collecting terms, this turns into a set of equations

$$\sum_{j} j^{\ell} L^{(t)}_{j} \mathbf{a}_{k} = \mathbf{b}_{k}, \quad k = 0, \ldots, \tilde{p} - 1, \tag{9.6}$$

where $\mathbf{a}_k$ and $\mathbf{b}_k$ are known vectors.

These equations can always be solved by using $\tilde{p}$ different coefficients $L^{(t)}_{j}$, $j = 0, \ldots, \tilde{p} - 1$. In most cases, fewer coefficients suffice.

Equations (9.6) are vector equations, and each $L_j^{(t)}$ is a matrix of size $r \times r$. In general, we can hope to satisfy r equations with each $L_j^{(t)}$, so we can try to get by with only $\lceil \tilde{p}/r \rceil$ nonzero coefficient matrices. (This is not always possible.) ∎

Example 9.2
A basic completion of the cubic Hermite splines to a biorthogonal multiwavelet has the symbols

$$\begin{pmatrix} H(z) \\ G(z) \end{pmatrix} = \frac{1}{16} \begin{pmatrix} 4+8z+4z^2 & 6-6z^2 \\ -1+z^2 & -1+4z-z^2 \\ 8 & 0 \\ 0 & 8 \end{pmatrix},$$

$$\begin{pmatrix} \tilde{H}(z) \\ \tilde{G}(z) \end{pmatrix} = \frac{1}{4z} \begin{pmatrix} 4z^2 & 0 \\ 0 & 8z^2 \\ -2+4z-2z^2 & -1+z^2 \\ 3-3z^2 & 1+4z+z^2 \end{pmatrix}.$$

H has approximation order 4, $\tilde{H}$ has approximation order 0.

We can raise the dual approximation order from 0 to 2 with a dual lifting step with constant matrix

$$L(z) = \frac{1}{4} \begin{pmatrix} -2 & 15 \\ 0 & -1 \end{pmatrix}. \tag{9.7}$$

This produces

$$\begin{pmatrix} H_{\text{new}}(z) \\ G_{\text{new}} \end{pmatrix} = \frac{1}{64} \begin{pmatrix} 16+32z+16z^2 & 24-24z^2 \\ -4+4z^2 & -4+16z-4z^2 \\ 9-16z+7z^2 & -27+60z-3z^2 \\ 1-z^2 & 33-4z+z^2 \end{pmatrix},$$

$$\begin{pmatrix} \tilde{H}_{\text{new}}(z) \\ \tilde{G}_{\text{new}} \end{pmatrix} = \frac{1}{16z} \begin{pmatrix} -4+8z+12z^2 & -2+2z^2 \\ 33-60z+27z^2 & 16+4z+18z^2 \\ -8+16z-8z^2 & -4+4z^2 \\ 12-12z^2 & 4+16z+4z^2 \end{pmatrix}.$$

More examples can be found in [98]. ∎

10

Creating Multiwavelets

We have seen some examples of multiscaling and multiwavelet functions in earlier chapters. In this chapter we will discuss general ways of modifying existing multiwavelets, or creating multiwavelets with desired properties from scratch.

Before that, we will discuss the *completion problem*. Given a multiscaling function ϕ, how do we find its dual $\tilde{\phi}$ and the multiwavelet functions?

10.1 Orthogonal Completion

Given an orthogonal multiscaling function, how do we find the multiwavelet functions? In the multiwavelet case it is not as easy as in the scalar case, but it is still not hard to do. We will present two different methods. In addition to providing algorithms, this shows that the completion problem can always be solved, and that the multiwavelets have compact support if ϕ has compact support.

The multiwavelets will not be unique. However, we have this theorem.

THEOREM 10.1
If $P_1(z)$ and $P_2(z)$ are two paraunitary polyphase matrices with the same multiscaling functions, then they are related by

$$P_1(z) = \begin{pmatrix} I & 0 \\ 0 & T(z) \end{pmatrix} P_2(z),$$

where $T(z)$ is paraunitary.

This is proved, for example, in [54].

10.1.1 Using Projection Factors

One way to solve the completion problem is based on projection factors. As noted before, it is not necessary to have the complete polyphase matrix $P(z)$

to do a decomposition into projection factors. The following is a corollary of theorem 9.2.

COROLLARY 10.2
If
$$P_H(z) = (H_0(z) H_1(z) \cdots H_{m-1}(z))$$
are the polyphase symbols of an orthogonal multiscaling function, we can find orthogonal projection factors $F_k(z)$ so that
$$P_H(z) = Q_H F_1(z) \cdots F_n(z),$$
where Q_H is an $r \times mr$ constant matrix with orthonormal rows:
$$Q_H Q_H^* = I.$$

The rest is easy: extend Q_H to a square unitary matrix Q (or even a square paraunitary matrix $Q(z)$); then
$$P(z) = Q F_1(z) \cdots F_n(z)$$
will be the desired extension to a full paraunitary polyphase matrix.

To find the projection factors, we need to choose
$$R(P_{H,n}^*) \subset R(U) \subset N(P_{H,0})$$
like before. Since $\dim R(P_{H,n}^*) \leq r$, $\dim N(P_{H,0}) \geq (m-1)r$, there may be a wide range of choices for U. The choice $R(U) = R(P_{H,n}^*)$ will keep the McMillan degree of the completion as low as possible.

Example 10.1
The polyphase symbols of the DGHM multiscaling function factor as
$$\begin{pmatrix} H_0(z) & H_1(z) \end{pmatrix} = Q F(z),$$
where
$$Q = \frac{1}{10} \begin{pmatrix} 3\sqrt{2} & 8 & 3\sqrt{2} & 0 \\ 4 & -3\sqrt{2} & 4 & 5\sqrt{2} \end{pmatrix},$$
and the projection factor F is based on
$$\mathbf{u} = \frac{1}{10} \begin{pmatrix} 9 \\ -3\sqrt{2} \\ -1 \\ 0 \end{pmatrix}.$$

We can complete Q in any orthogonal fashion, and obtain the multiwavelet functions. □

10.1.2 Householder-Type Approach

An ingenious method is described in [103]. I think of the following theorem as a matrix polynomial version of the construction of Householder matrices.

THEOREM 10.3
Assume $\mathbf{a}(z)$ is an orthonormal trigonometric vector polynomial; that is,
$$\mathbf{a}^*(z)\mathbf{a}(z) = 1.$$
Then there exists a paraunitary matrix $U(z)$ so that
$$U(z)\mathbf{a}(z) = (1, 0, \ldots, 0)^*.$$

PROOF We will construct $U(z)$ as a product of several paraunitary matrices: Householder matrices and diagonal matrices of the form

$$D(z) = \begin{pmatrix} z^{-d_1} & & & \\ & z^{-d_2} & & \\ & & z^{-d_3} & \\ & & & \ddots \end{pmatrix}. \tag{10.1}$$

Given $\mathbf{a}^{(0)} = \mathbf{a}$, we choose a diagonal matrix $D_1(z)$ so that each nonvanishing entry in $D_1(z)\mathbf{a}^{(0)}(z)$ begins with a nonzero constant term. That is,
$$D_1(z)\mathbf{a}^{(0)}(z) = \mathbf{a}_0^{(1)} + \mathbf{a}_1^{(1)}z + \cdots + \mathbf{a}_n^{(1)}z^n,$$
with all entries in $\mathbf{a}_0^{(1)}$ nonzero (unless the corresponding entry in $\mathbf{a}^{(0)}(z)$ is identically zero).

If $n \geq 1$, the orthonormality of $\mathbf{a}$ implies that
$$\mathbf{a}_0^{(1)*}\mathbf{a}_n^{(1)} = 0. \tag{10.2}$$
We choose a Householder matrix U_1 so that
$$U_1 \mathbf{a}_0^{(1)} = \begin{pmatrix} a_{0,1}^{(2)} \\ 0 \\ \vdots \\ 0 \end{pmatrix}.$$

Multiplication by U_1 preserves equation (10.2), so
$$U_1 \mathbf{a}_n^{(1)} = \begin{pmatrix} 0 \\ a_{n,2}^{(2)} \\ a_{n,3}^{(2)} \\ \vdots \end{pmatrix}.$$

We can apply another diagonal matrix $D_2(z)$ to reduce the polynomial degree of each component except the first, and find that

$$\mathbf{a}^{(2)} = D_2(z)U_1\mathbf{a}^{(1)}$$

is still orthonormal and has degree at most $n-1$. Eventually, we get to

$$U_n D_n(z) U_{n-1} \cdots U_1 D_1(z) \mathbf{a}(z) = (1, 0, 0, \ldots)^*.$$ ∎

COROLLARY 10.4
If $A(z)$ is a trigonometric matrix polynomial with orthonormal columns, we can find a paraunitary matrix $U(z)$ so that

$$U(z)A(z) = \begin{pmatrix} I \\ 0 \end{pmatrix}.$$

This is just a recursive application of theorem 10.3.

Example 10.2
For the DGHM multiwavelet, the multiscaling function part of the polyphase matrix can be reduced by

$$Q(z) = \begin{pmatrix} 0 & 0.4500 & -0.7124 & -0.3265 \\ 0 & -0.2121 & 0.1399 & 0.4852 \\ 0 & -0.0500 & 0.4487 & -0.5884 \\ 0 & 0 & -0.2939 & 0.4969 \end{pmatrix} z^{-1} + \begin{pmatrix} 0.4243 & -0.0500 & -0.0248 & -0.0147 \\ 0.8000 & -0.2121 & -0.1054 & -0.0624 \\ 0.4243 & 0.4500 & 0.2236 & 0.1323 \\ 0 & 0.7071 & 0.3514 & 0.2078 \end{pmatrix}$$

to the form

$$\begin{pmatrix} H_0(z) & H_1(z) \end{pmatrix} Q(z) = \begin{pmatrix} 1 & 0 & 0 & 0 \\ 0 & 1 & 0 & 0 \end{pmatrix}.$$

We can then complete the matrix on the right, for example, to the identity, and multiply by Q^* to obtain a full $P(z)$. □

10.2 Biorthogonal Completion

The paper [103] cited above contains a second algorithm useful for biorthogonal multiwavelets. I think of the following theorem as a polynomial version of QR-factorization.

THEOREM 10.5
Assume $\mathbf{a}(z)$ is a trigonometric vector polynomial. Then there exists an invertible trigonometric matrix polynomial $T(z)$ so that

$$T(z)\mathbf{a}(z) = (p(z), 0, \ldots, 0)^*,$$

Chapter 10: Creating Multiwavelets

where $p(z)$ is a scalar polynomial.

PROOF The construction is similar to the orthogonal case, but it requires elementary matrices as building blocks, in addition to Householder and diagonal matrices. (An elementary matrix is an identity matrix with off-diagonal terms only in a single row or column.) ∎

REMARK 10.6 The relationships

$$T(z)\mathbf{a}(z) = (p(z), 0, \ldots, 0)^*,$$
$$\mathbf{a}(z) = T(z)^{-1}(p(z), 0, \ldots, 0)^*,$$

show that $\mathbf{a}(z)$ vanishes if and only if $p(z)$ vanishes; thus, $p(z)$ is the greatest common divisor (gcd) of the elements of $\mathbf{a}(z)$. ∎

By using theorem 10.5 recursively, we obtain the following corollary.

COROLLARY 10.7

If $A(z)$ is a trigonometric matrix polynomial of size $s \times t$, we can find an invertible matrix polynomial $T(z)$ and a permutation matrix E so that

$$T(z)A(z)E = \begin{pmatrix} B_{11}(z) & B_{12}(z) \\ 0 & 0 \end{pmatrix},$$

where

$$B_{11}(z) = \begin{pmatrix} b_{11}(z) & b_{12}(z) & \cdots & b_{1s}(z) \\ 0 & b_{22}(z) & \ddots & \vdots \\ \vdots & \ddots & \ddots & \vdots \\ 0 & \cdots & 0 & b_{ss}(z) \end{pmatrix}, \tag{10.3}$$

with $s = \operatorname{rank}(A(z))$, and each $b_{kk}(z)$ not identically zero.

Alternatively, we could find an invertible matrix polynomial $T(z)$ so that

$$T(z)A(z) = \begin{pmatrix} B_{11}(z) & B_{12}(z) \end{pmatrix} \qquad \text{if } s \leq t$$

or

$$T(z)A(z) = \begin{pmatrix} B_{11}(z) \\ 0 \end{pmatrix} \qquad \text{if } s \geq t,$$

where $B_{11}(z)$ has the form in equation (10.3), but possibly with some zero elements on the diagonal.

This approach can solve the completion problem in the biorthogonal case.

COROLLARY 10.8
If
$$P_H(z) = (H_0(z) H_1(z) \cdots H_{m-1}(z))$$
are the polyphase symbols of a biorthogonal multiscaling function, we can find an invertible matrix $T(z)$ so that
$$P_H(z) T(z) = (B_0(z)\ 0\ \cdots\ 0) \quad (10.4)$$
where $B_0(z)$ has the form in equation (10.3).

ϕ has a dual if and only if the diagonal terms of $B(z)$ are nonzero monomials. If this is the case, the complete polyphase matrix must be of the form

$$P(z) = \begin{pmatrix} \begin{array}{c|c} B_0(z) & 0 \\ \hline C_1(z) & \\ \vdots & D(z) \\ C_{m-1}(z) & \end{array} \end{pmatrix} T(z)^{-1}$$

where $D(z)$ is any invertible matrix polynomial, and $C_k(z)$ are arbitrary.

Example 10.3
For the cubic Hermite multiscaling function, the polyphase matrix can be reduced to
$$(H_0(z)\ H_1(z))\, T(z) = \begin{pmatrix} 1.2051 & 0 & 0 & 0 \\ -0.0079z^{-1} - 0.1084 + 0.0277z & 0.4228 & 0 & 0 \end{pmatrix},$$
by
$$T(z) = \begin{pmatrix} 0 & 0.0269 & -0.8988 & 0 \\ 0 & -0.0157 & 0.5228 & 0 \\ 0 & -0.0017 & 0.0573 & 0 \\ 0 & 0.0028 & -0.0940 & 0 \end{pmatrix} z^{-2} + \begin{pmatrix} 0.2144 & 2.6345 & -0.1384 & -0.5596 \\ -0.1247 & -1.5265 & -0.1198 & 0.3256 \\ -0.0137 & -0.1976 & 1.0006 & 0.0357 \\ 0 & 0.2664 & 0.2909 & -0.0585 \end{pmatrix} z^{-1}$$
$$+ \begin{pmatrix} 0.9393 & 0.1872 & 0.0056 & 0.2663 \\ 0.6262 & 0.1872 & 0.0056 & 0.2663 \\ 0.5643 & -2.6962 & -0.0276 & 0.1911 \\ 0 & 0.2492 & 0.0075 & 0.3545 \end{pmatrix} + \begin{pmatrix} 0 & 0 & 0 & 0 \\ 0 & 0 & 0 & 0 \\ 0 & 0.0468 & 0.0014 & 0.0666 \\ 0 & 0 & 0 & 0 \end{pmatrix} z$$

We can complete the 2×4 matrix on the right to any invertible matrix, and then multiply on the right by T^{-1}. □

10.3 Other Approaches

The **cofactor method** is described in [135]. It can be used to find the dual multiscaling functions. In the case $m = 2$, it can also be adapted to find the multiwavelet functions.

Chapter 10: Creating Multiwavelets

Let $K(z)$ be the cofactor matrix of the symbol $H(z)$. Then
$$H(z)K(z) = \Delta(z) \cdot I, \qquad \Delta(z) = \det(H(z)).$$
If we can find a scalar polynomial $\tilde{\Delta}(z)$ so that
$$\sum_{k=0}^{m-1} \Delta(w^k z)\tilde{\Delta}(w^k z)^* = 1, \qquad w = e^{2\pi/m}, \tag{10.5}$$
then
$$\tilde{H}(z) = \tilde{\Delta}(z) K(z)^*$$
will satisfy the biorthogonality conditions. $\tilde{\Delta}$ can be found using the method of theorem 10.5.

The paper [68] describes an algorithm for finding the multiwavelet functions if ϕ, $\tilde{\phi}$ are already known. It is similar to the QR-type approach.

10.4 Techniques for Modifying Multiwavelets

A technique for modifying multiwavelets is one which creates new multiwavelets ϕ_{new}, $\psi^{(t)}_{\text{new}}$, $\tilde{\phi}_{\text{new}}$, $\tilde{\psi}^{(t)}_{\text{new}}$ by applying some transformation to given ϕ, $\psi^{(t)}$, $\tilde{\phi}$, $\tilde{\psi}^{(t)}$.

Some examples are

- **Applying a TST.** This has the effect of shifting approximation orders from one side to the other, similar to moving a factor of $(1 + e^{-i\xi})/2$ in the scalar case.

  Symmetry can be preserved or even created. Orthogonality is destroyed. In general, both ϕ and $\tilde{\phi}$ get longer, but adding symmetry conditions helps to keep the length increase under control.

  Examples for this technique are found in chapter 8 and in [121], [135], and [136].

- **Using projection factors.** We can add a projection factor to an existing multiwavelet. This method has the advantage that it preserves orthogonality if F is orthogonal. Unfortunately, it also destroys approximation order beyond $p = 1$.

- **Using lifting factors.** We can apply a lifting factor or dual lifting factor to an existing multiwavelet. Appropriately chosen lifting factors can preserve or increase approximation order, but they destroy orthogonality.

  Examples are found in chapter 9 and in [54] and [98].

10.5 Techniques for Building Multiwavelets

A technique for building multiwavelets is one which creates multiwavelets from scratch. The standard way to go about this is to begin with one of the known factorizations of the symbol or polyphase matrix with some free parameters in the factors, add other desired conditions, and try to solve it using a computer algebra system, or by a numerical method.

- **Using the TST factorization.** We set up

$$H(\xi) = m^{-p} C_p(m\xi) \cdots C_1(m\xi) H_0(\xi) C_1(\xi)^{-1} \cdots C_p(\xi)^{-1}.$$

This TST factorization will automatically provide approximation order, even balanced approximation order p, but orthogonality must be added as a constraint. This approach is similar in spirit to the original construction of the Daubechies wavelets.

The TST method is not seen very often, but Lebrun and Vetterli [104] have constructed orthogonal symmetric balanced multiwavelets of order 1, 2, and 3 this way. The resulting multiwavelets are called BAT O1, BAT O2, and BAT O3. The coefficients for BAT O1 are listed in appendix A.

- **Using projection factors.** We set up

$$P(\xi) = Q F_1(\xi) \cdots F_k(\xi).$$

This is an easy way to create orthogonal multiwavelets with approximation order 1 of any size. Higher approximation orders and symmetry need to be imposed as extra constraints.

Examples of this approach for scalar wavelets are presented in sections 3.6.2 and 3.7.2.

Hardin and Roach used this method to construct their prefilters [76]. The Shen–Tan–Tam multiwavelets [130] were also constructed from projection factors.

- **Using lifting factors.** We set up

$$P(z) = L_k(z) \cdots L_1(z) P_0(z),$$

where the L_j are lifting factors or dual lifting factors, and P_0 is a simple initial polyphase matrix. P_0 could be an identity matrix.

This is an easy way to create biorthogonal multiwavelets of any size. Approximation order is easy to enforce, even balanced approximation order, and it is possible to create symmetric multiwavelet this way.

Examples can be found in [12], [13], [54], [67], and [145].

- **Other approaches.** The DGHM multiwavelet [59] was constructed by a completely different procedure, using fractal interpolation functions. The Hardin–Marasovich multiwavelets [75] are also based on this idea.

  The Chui–Lian multiwavelets [36] were derived by solving the orthogonality conditions directly, after adding symmetry and interpolation constraints. I believe it would be hard to push this brute force approach beyond what was done in the original paper.

  A new type of multiwavelet construction method is described in [73] and [74].

11

Existence and Regularity

Previous chapters explained the basic ideas behind refinable function vectors, multiresolution approximations (MRAs) and the discrete multiwavelet transform (DMWT), as well as ways for determining some basic properties of the basis functions: approximation order, moments, and point values.

These results were all presented under the assumption that the underlying refinement equation defines a multiscaling function ϕ with some minimal regularity properties, and that this function produces an MRA with multiwavelet functions $\psi^{(t)}$. In the biorthogonal case, we also assumed the existence of $\tilde{\phi}$ and $\tilde{\psi}^{(t)}$.

In order for the DMWT algorithm to work, it is actually not necessary that such functions really exist: if we have sets of recursion coefficients which satisfy the biorthogonality conditions in equation (6.15), they will give rise to a DMWT algorithm that works on a purely algebraic level. We may not be able to justify the interpretation of the DMWT as a splitting of the original signal into a coarser approximation and fine detail at different levels; there may also be numerical stability problems when we decompose and reconstruct over many levels, but the algorithm will be an invertible transform.

In this chapter, we will give necessary and sufficient conditions for existence, regularity, and stability of ϕ and $\psi^{(t)}$. The material is rather mathematical.

We note that to establish existence and regularity, it suffices to look at the multiscaling function ϕ. The multiwavelet functions $\psi^{(t)}$ are just finite linear combinations of scaled translates of ϕ, so they automatically inherit those properties. We only need to look at $\psi^{(t)}$ to check their stability.

There are two main approaches: in the time domain (section 11.4) and in the frequency domain (sections 11.1 to 11.3.) The time domain approach is based on the refinement equation

$$\phi(x) = \sqrt{m} \sum_{k=k_0}^{k_1} H_k \, \phi(mx - k); \qquad (11.1)$$

the frequency domain approach is based on the Fourier transform of equation (11.1), which is

$$\hat{\phi}(\xi) = H(\xi/m)\hat{\phi}(\xi/m), \qquad (11.2)$$

where $H(\xi)$ is the symbol of ϕ, defined by

$$H(\xi) = \frac{1}{\sqrt{m}} \sum_{k=k_0}^{k_1} H_k e^{-ik\xi}.$$

11.1 Distribution Theory

Ultimately we are interested in solutions of the refinement equation (11.1), but at first we will consider solutions of the Fourier-transformed equation (11.2). We are looking for solutions for which $\hat{\phi}$ is a function vector, but ϕ itself could be a vector of distributions.

Recall the definition of condition E(p) (definition 6.8).

THEOREM 11.1
A necessary condition for the existence of a function vector $\hat{\phi}(\xi)$ which is continuous at 0 with $\hat{\phi}(0) \neq \mathbf{0}$ and which satisfies equation (11.2) is that $H(0)$ has an eigenvalue of 1 with eigenvector $\hat{\phi}(0)$.

A sufficient condition is that $H(0)$ satisfies condition E(p) for some p. If it does, then

- *The product*

$$\Pi_n(\xi) = \prod_{k=1}^{n} H(2^{-k}\xi) \qquad (11.3)$$

  converges uniformly on compact sets to a continuous matrix-valued limit function Π_∞ with polynomial growth.

- *For any nonzero choice of $\hat{\phi}(0)$ with $H(0)\hat{\phi}(0) = \hat{\phi}(0)$,*

$$\hat{\phi}(\xi) = \Pi_\infty(\xi)\hat{\phi}(0)$$

  satisfies equation (11.2), and is the Fourier transform of a tempered distribution vector ϕ with compact support in the interval $[k_0/(m-1), k_1/(m-1)]$.

- *There are exactly p linearly independent solutions $\hat{\phi}_1, \ldots, \hat{\phi}_p$, corresponding to p linearly independent eigenvectors $\hat{\phi}_1(0), \ldots, \hat{\phi}_p(0)$ of $H(0)$ to eigenvalue 1.*

PROOF (See [79].) The necessity is easy. For $\xi = 0$, equation (11.2) says

$$\hat{\phi}(0) = H(0)\hat{\phi}(0).$$

Chapter 11: Existence and Regularity

If $\hat{\boldsymbol{\phi}}(0) \neq \mathbf{0}$, it must be an eigenvector to eigenvalue 1 of $H(0)$.

Assume now that $H(0)$ satisfies condition E(p). This implies that we can choose a matrix norm for which

$$\|H(0)\| = 1.$$

Here is where we need the fact that the eigenvalue 1 is nondegenerate.

The rest of the proof is identical to the proof in the scalar case (theorem 5.1), except that now we have multiple solutions. ∎

REMARK 11.2 The preceding proof shows that the given solutions $\boldsymbol{\phi}_k$, $k = 1, \ldots, p$ are the only ones whose Fourier transforms are function vectors continuous at 0 with $\hat{\boldsymbol{\phi}}_k(0) \neq \mathbf{0}$. The uniqueness disappears if you drop any of these assumptions.

There may be other distribution solutions with $\hat{\boldsymbol{\phi}}(0) = \mathbf{0}$, there may be solutions with discontinuous $\hat{\boldsymbol{\phi}}$, and there may even be distribution solutions which have no Fourier transform. For example, the Hilbert transform of any tempered distribution solution is also a solution. ∎

Theorem 11.1 can be generalized.

THEOREM 11.3
A necessary and sufficient condition for the existence of a compactly supported distribution solution of the refinement equation is that $H(0)$ has an eigenvalue of the form m^n for some integer $n \geq 0$.

If $n > 0$, then $\hat{\boldsymbol{\phi}}(0) = \mathbf{0}$, and $\boldsymbol{\phi}$ is the nth derivative of a distribution solution $\boldsymbol{\Phi}(\xi)$ of

$$\boldsymbol{\Phi}(x) = m^n \sqrt{m} \sum_{k=a}^{b} H_k \boldsymbol{\Phi}(mx - k)$$

with $\hat{\boldsymbol{\Phi}}(0) \neq \mathbf{0}$.

The necessity is proved in [79]. Sufficiency is proved in [93]
Let

$$\rho = \rho(H(0))$$

be the spectral radius of $H(0)$, that is, the magnitude of the largest eigenvalue of $H(0)$.

It seems reasonable to assume that we need $\rho = 1$ if we expect solutions with continuous $\hat{\boldsymbol{\phi}}$, but that is not the case.

The following theorem says that we still get solutions if $1 < \rho < m$, and there are examples that show that even for $\rho \geq m$ continuous solutions sometimes exist. (Always assuming that $H(0)$ has an eigenvalue of 1, of course.)

THEOREM 11.4
If $\rho = \rho(H(0)) < m$, then for any vector $\hat{\phi}(0)$ with $H(0)\hat{\phi}(0) = \hat{\phi}(0)$, the product
$$\phi_n(\xi) = \Pi_n(\xi)\hat{\phi}(0)$$
converges uniformly on compact sets to a continuous limit function
$$\hat{\phi}(\xi) = \Pi_\infty \hat{\phi}(0)$$
which satisfies equation (11.2).

PROOF Choose $\epsilon > 0$ and a matrix norm so that
$$\|H(0)\| = \rho + \epsilon < m.$$
Let
$$q = \frac{\|H(0)\|}{m} < 1.$$
$H(\xi)$ is 2π-periodic and differentiable, so we can find constants c and α so that for all ξ
$$\|H(\xi) - H(0)\| \leq c\|H(0)\| |\xi|$$
$$\|H(\xi)\| \leq m^\alpha.$$

The first of these inequalities implies
$$\|H(\xi)\| \leq \|H(\xi) - H(0)\| + \|H(0)\| \leq \|H(0)\|(1 + c|\xi|) \leq \|H(0)\|e^{c|\xi|}.$$
Then
$$\begin{aligned}\|\Pi_n(\xi)\| &\leq \|H(0)\|^n e^{c|\xi|(m^{-1} + \cdots m^{-n})} \\ &< \|H(0)\|^n e^{c|\xi|/(m-1)} \leq \|H(0)\|^n e^{c|\xi|}.\end{aligned} \quad (11.4)$$

For any ξ,
$$\begin{aligned}\left\|(\Pi_n(\xi) - \Pi_{n-1}(\xi))\hat{\phi}(0)\right\| &= \left\|\Pi_{n-1}(\xi)\left[H(m^{-n}\xi) - H(0)\right]\hat{\phi}(0)\right\| \\ &\leq \|H(0)\|^{n-1}e^{c|\xi|} \cdot c\|H(0)\|m^{-n}|\xi| \cdot \|\hat{\phi}(0)\| \\ &= \left(\frac{\|H(0)\|}{m}\right)^n c|\xi|e^{c|\xi|}\|\hat{\phi}(0)\| \\ &= q^n \cdot \left(c|\xi|e^{c|\xi|}\|\hat{\phi}(0)\|\right).\end{aligned}$$

For any $m > n$, by a telescoping series argument
$$\left\|(\Pi_m(\xi) - \Pi_n(\xi))\hat{\phi}(0)\right\| \leq \frac{q^n}{1-q}\left(c|\xi|e^{c|\xi|}\|\hat{\phi}(0)\|\right).$$

Chapter 11: Existence and Regularity 223

This shows that $\Pi_n(\xi)\hat{\phi}(0)$ converges uniformly as $n \to \infty$ for ξ in any compact set. ∎

This proof does not show that $\hat{\phi}$ has polynomial growth. In [79], this theorem is proved by a different method under the additional assumption that $H(0)$ has a single nondegenerate eigenvalue λ with $1 < |\lambda| < m$. In that case you do get polynomial growth, so ϕ is a distribution solution with compact support. The authors call this a *constrained solution*.

A constrained solution has the property that $\Pi_n(\xi)\hat{\phi}(0)$ converges, even though $\Pi_n(\xi)$ does not. There may also be solutions where not even $\Pi_n(\xi)\hat{\phi}(0)$ converges. These are called *superconstrained solutions* in [79].

Example 11.1
If we multiply the coefficients of the DGHM multiscaling function example by -5, we find that $H(0)$ has eigenvalues 1 and -5, but a superconstrained solution exists:
$$\phi = \begin{pmatrix} \sqrt{2}\chi_{[0,1)} \\ -\chi_{[0,2)} \end{pmatrix}.$$
□

11.2 L^1-Theory

In this section we will look for sufficient conditions for the existence of an L^1-solution of the refinement equation.

We already have some sufficient conditions that guarantee the existence of a continuous $\hat{\phi}$. To show that this function has an inverse Fourier transform which is a function, we need to impose decay conditions on $|\hat{\phi}(\xi)|$ as $|\xi| \to \infty$. As in the scalar case, this requires approximation order conditions.

Recall the definition of C^α (definition 5.4).

THEOREM 11.5
Assume the symbol $H(\xi)$ satisfies the sum rules of order p for some $p \geq 1$, so it factors as
$$H(\xi) = m^{-p} C_p(m\xi) \cdots C_1(m\xi) H_0(\xi) C_1(\xi)^{-1} \cdots C_p(\xi)^{-1}.$$
If $\rho(H_0(0)) < m$ and
$$\sup_\xi \|H_0(\xi)\| < m^{p-\alpha-1},$$
then $\phi \in C^\alpha$.

More generally, we can replace the bound by
$$\sup_\xi \|\Pi_{0,n}(\xi)\|^{1/n} < m^{p-\alpha-1}$$

for some n, where
$$\Pi_{0,n}(\xi) = \prod_{k=1}^{n} H_0(2^{-k}\xi)$$

PROOF We will just mention some of the key steps. The full proof is given in [41] and [122]. It is quite technical, and the theorem does not give very useful estimates in practice. The L^2-estimates below are much more practical.

We want to show that
$$\|\Pi_\infty(\xi)\| \leq c(1+|\xi|)^{-\alpha-1-\epsilon} \tag{11.5}$$
for some $\epsilon > 0$. For $|\xi| < 1$, everything is bounded, so we can assume $|\xi| \geq 1$.

$\Pi_n(\xi)$ is a telescoping product: when we expand it, all C_k-terms on the inside cancel, so
$$\Pi_n(\xi) = \frac{1}{m^{np}} C_p(m\xi) \cdots C_1(m\xi) \Pi_{0,n}(\xi) C_1(m^{-n}\xi)^{-1} \cdots C_p(m^{-n}\xi)^{-1}.$$

Using the definition of C_0 from lemma 8.8, we get
$$\Pi_n(\xi) = \left(\frac{1}{m^n(1-e^{-im^{-n}\xi})}\right)^p C_p(m\xi) \cdots C_1(m\xi)$$
$$\times \Pi_{0,n}(\xi) C_{1,0}(m^{-n}\xi) \cdots C_{p,0}(m^{-n}\xi).$$

The first term becomes
$$\lim_{n\to\infty} \left(\frac{1}{m^n(1-e^{-im^{-n}\xi})}\right)^p = |\xi|^{-p} \leq 2(1+|\xi|)^{-p}.$$

The product of $C_k(m\xi)$ is uniformly bounded.

For the remaining terms, we make the simplifying assumption that the two-scale similarity transform (TST) matrices C_k have the standard form given in example 8.1
$$C_k(\xi) = I - \mathbf{r}_k \mathbf{l}_k^* e^{-i\xi}.$$
Next, we establish that we can replace $C_{k,0}(m^{-n}\xi)$ by $C_{k,0}(0)$ (see [41] and [122] for details). This sets the stage for the main trick: for any vector $\mathbf{a}$,
$$C_{k,0}(0)\mathbf{a} = \mathbf{r}_k \mathbf{l}_k^* \mathbf{a}$$
is a multiple of $\mathbf{r}_k$. Thus,
$$\Pi_{0,n} C_{1,0}(0) \cdots C_{p,0}(0)\mathbf{a} = (\mathbf{l}_1^* \mathbf{r}_2) \cdots (\mathbf{l}_{p-1}^* \mathbf{r}_p)(\mathbf{l}_p^* \mathbf{a}) \Pi_{0,n} \mathbf{r}_1$$
$$\|\Pi_{0,n}(\xi) C_{1,0}(0) \cdots C_{p,0}(0)\mathbf{a}\| \leq c \|\Pi_{0,n}(\xi)\mathbf{r}_1\| \|\mathbf{a}\|,$$
$$\|\Pi_{0,n}(\xi) C_{1,0}(0) \cdots C_{p,0}(0)\| \leq c \|\Pi_{0,n}(\xi)\mathbf{r}_1\|.$$

$\mathbf{r}_1$ an eigenvector of $H(0)$ to eigenvalue 1 by theorem 8.4, so we can use the constrained convergence estimates from theorem 11.4 to show that
$$\|\Pi_{0,n}(\xi) C_{1,0}(0) \cdots C_{p,0}(0)\| \leq c(1+|\xi|)^{p-\alpha-1-\epsilon}. \qquad \blacksquare$$

11.3 L^2-Theory

11.3.1 Transition Operator

DEFINITION 11.6 *The* transition operator *or* transfer operator *for the symbol $H(\xi)$ is defined by*

$$TF(\xi) = \sum_{k=0}^{m-1} H(\xi + \frac{2\pi}{m}k)F(\xi + \frac{2\pi}{m}k)H(\xi + \frac{2\pi}{m}k)^*. \quad (11.6)$$

The transition operator maps 2π-periodic matrix-valued functions into 2π-periodic matrix-valued functions, but it also maps some smaller spaces into themselves.

Let E_n be the space of trigonometric matrix polynomials of the form

$$A(\xi) = \sum_{k=-n}^{n} A_k e^{-ik\xi};$$

and let F_n be the subspace of those $A \in E_n$ with $\mathbf{y}_0^* A(0) = 0$, where $\mathbf{y}_0 \neq \mathbf{0}$ is the zeroth approximation vector of H.

LEMMA 11.7
If $H(\xi)$ has degree N, the transition operator T maps E_n into itself for any $n > (N-m)/(m-1)$.

If H satisfies the sum rules of order 1, then T also maps F_n into itself for any $n > (N-m)/(m-1)$.

The proof is the same as that in the scalar case.

We denote the restriction of T to E_n by T_n.

The following lemma from [107] is the main trick which will allow us to get better decay estimates for $|\hat{\phi}(\xi)|$.

LEMMA 11.8
If T is the transition operator for $H(\xi)$, then for any 2π-periodic L^2-matrix function $A(\xi)$

$$\int_{-m^k\pi}^{m^k\pi} \Pi_k(\xi) A(m^{-k}\xi) \Pi_k(\xi)^* \, d\xi = \int_{-\pi}^{\pi} T^k A(\xi) \, d\xi.$$

Where does the transition operator come from? If $\phi \in L^2$, we can define the function

$$A(x) = \langle \phi(y), \phi(y-x) \rangle = \int \phi(y) \phi(y-x)^* \, dy.$$

Its Fourier transform is given by

$$\hat{A}(\xi) = \sqrt{2\pi} \hat{\phi}(\xi) \hat{\phi}(\xi)^*,$$

which makes it useful for studying the L^2-properties of ϕ. However, this is not quite the function we want to use.

DEFINITION 11.9 The autocorrelation matrix of $\phi \in L^2$ is defined as

$$\Omega(\xi) = \sum_k A(k) e^{-ik\xi}. \qquad (11.7)$$

If ϕ has compact support, only finitely many of the $A(k)$ will be nonzero, so Ω is a trigonometric matrix polynomial. It is easier to compute and easier to work with than $\hat{A}(\xi)$.

By using the Poisson summation formula, we can verify that

$$\Omega(\xi) = \sqrt{2\pi} \sum_k \hat{\phi}(\xi + 2\pi k) \hat{\phi}(\xi + 2\pi k)^*,$$

so

$$\int_0^{2\pi} \Omega(\xi) \, d\xi = \sqrt{2\pi} \int_{\mathbb{R}} |\hat{\phi}(\xi)|^2 \, d\xi.$$

In the cascade algorithm, let $\Omega^{(n)}$ be the autocorrelation function of $\phi^{(n)}$. We can then verify that

$$\Omega^{(n+1)} = T \Omega^{(n)},$$

so the properties of the transition operator are intimately related with the convergence of $\Omega^{(n)}$, which in turn is related to the L^2-convergence of $\phi^{(n)}$.

For starters, we see that T must have an eigenvalue of 1, with the autocorrelation matrix as the corresponding eigenmatrix.

For practical computations, we can work with matrices instead of operators.

DEFINITION 11.10 The transition matrix or transfer matrix T_n is defined by

$$(T_n)_{k\ell} = \sum_s \overline{H_{s+\ell-2k}} \otimes H_s, \qquad -n \le k, \ell \le n. \qquad (11.8)$$

Here $\otimes$ stands for the Kronecker product, and $\overline{H_k}$ is the element-wise complex conjugate of the matrix H_k (not the complex conjugate transpose).

Chapter 11: Existence and Regularity   227

Where does this matrix come from?

The matrix-valued function $A(x)$ is refinable. When we substitute the recursion relation for ϕ into the definition of $A(x)$, we get

$$A(x) = \sum_{nk} H_k A(mx - n) H_{k-n}^*.$$

We string the matrix A out into a vector $\text{vec}(A)$, column by column.

The relationship
$$\text{vec}(ABC) = (C^T \otimes A)\text{vec}(B)$$
then produces
$$\text{vec}(A(x)) = \sqrt{m} \sum_n \mathbf{H}_n \text{vec}(A(mx-n)),$$

where
$$\mathbf{H}_n = \frac{1}{\sqrt{m}} \sum_k \overline{H_{k-n}} \otimes H_k.$$

The matrix T_n is the matrix T from section 6.5, for determining the point values of $A(x)$ at the integers.

There is a close relationship between the transition matrix and the transition operator, which is why we use the same notation for both.

We identify the matrix trigonometric polynomial $A(\xi) = \sum_k A_k e^{-ik\xi} \in E_n$ with the vector
$$\mathbf{A} = \begin{pmatrix} \text{vec}(A_{-n}) \\ \cdots \\ \text{vec}(A_n) \end{pmatrix} \in \mathbb{C}^{(2n+1)r^2}.$$

The vector and the function have equivalent norms
$$\int_{-\pi}^{\pi} \|A(\xi)\|_F^2 \, d\xi = 2\pi \|\mathbf{A}\|_2^2,$$

where $\|\cdot\|_F$ is the Frobenius norm.

For $n \geq N - 1$, the matrix trigonometric polynomial $T_n A \in E_n$ (where T_n is the transition operator) then corresponds to the vector $T_n \mathbf{A} \in \mathbb{C}^{(2n+1)r^2}$ (where T_n is the transition matrix). We can switch back and forth between the two viewpoints. In particular, this gives us an easy way to compute eigenvalues and eigenvectors of T_n.

11.3.2 Sobolev Space Estimates

Recall the definition of the Sobolev space H_s (definition 5.11).

In the scalar setting we were able to show that the Sobolev norm estimates did not depend on whether we factored out the approximation orders. I am not completely sure that this is true in the multiwavelet case, but it makes no practical difference. We will prove the counterpart to theorem 5.12 without factoring out the approximation orders, and still get useful results by restricting the transition operator to a suitably small subspace.

In the following, we need to use some properties of positive semidefinite matrices. They are listed in appendix B.7. The notation $\Gamma(\xi) \geq 0$ means that $\Gamma(\xi)$ is positive semidefinite for all ξ, not that it has nonnegative entries.

THEOREM 11.11
Assume that $H(0)$ satisfies condition $E(p)$ for some p.
Choose a trigonometric matrix polynomial Γ with the following properties:

- $\Gamma(\xi)$ *is Hermitian for all ξ.*

- $\Gamma(\xi) \geq 0 \quad$ *for all ξ.*

- $\Gamma(\xi) \geq cI, \quad c > 0, \quad$ *for $\pi/2 \leq |\xi| \leq \pi$.*

Let ρ be the spectral radius of the transition operator T for H_0, restricted to the smallest T-invariant subspace that contains the function $\Gamma(\xi)$. Then $\phi \in H_s$ for any $s < -\log_4 \rho$.

PROOF The proof given here is based on [92]. It is quite similar to the proof in the scalar case.

Condition $E(p)$ implies that $\hat{\phi}(\xi)$ is bounded on $[-\pi, \pi]$.

Assume that we have chosen Γ and determined the smallest T-invariant subspace F that contains Γ. We choose $\epsilon > 0$ and a matrix norm so that

$$\|T^n A\| \leq (\rho + \epsilon)^n \|A\|$$

for all $A \in F$.

Let $\Omega_n = [-m^n \pi, -m^{n-1}\pi] \cup [m^{n-1}\pi, m^n \pi]$. Then

$$\int_{\Omega_n} \hat{\phi}(\xi) \hat{\phi}(\xi)^* \, d\xi = \int_{\Omega_n} \Pi_n(\xi) \hat{\phi}(m^{-n}\xi) \hat{\phi}(m^{-n}\xi)^* \Pi_n(\xi)^*$$

$$\leq c_1 \int_{\Omega_n} \Pi_n(\xi) \Pi_n(\xi)^* \, d\xi$$

$$\leq c_2 \int_{\Omega_n} \Pi_n(\xi) \Gamma(m^{-n}\xi) \Pi_n(\xi)^* \, d\xi$$

$$\leq c_2 \int_{-m^n}^{m^n} \Pi_n(\xi) \Gamma(m^{-n}\xi) \Pi_n(\xi)^* \, d\xi$$

$$= c_2 \int_{-\pi}^{\pi} T^n \Gamma(\xi) \, d\xi.$$

Chapter 11: Existence and Regularity

Then

$$\int_{\Omega_n} \sum_{k=1}^r |\hat{\phi}_k(\xi)|^2 \, d\xi = \int_{\Omega_n} \operatorname{trace}(\hat{\phi}(\xi)\hat{\phi}(\xi)^*) \, d\xi$$
$$\leq \int_{-\pi}^{\pi} \operatorname{trace}(T^n \Gamma(\xi)) \, d\xi$$
$$\leq c_3 \, \|T^n \Gamma\| \leq c_4 (\rho + \epsilon)^n \|\Gamma\|.$$

The rest is the same as before:

$$\int_{-\infty}^{\infty} (1 + |\xi|)^{2s} \sum_{k=1}^r |\hat{\phi}_k(\xi)|^2 \, d\xi$$
$$= \left(\int_{-\pi}^{\pi} + \sum_{n=1}^{\infty} \int_{\Omega_n} \right) (1 + |\xi|)^{2s} \sum_{k=1}^r |\hat{\phi}_k(\xi)|^2 \, d\xi$$
$$\leq c + c \sum_{n=1}^{\infty} 2^{2ns} (\rho + \epsilon)^n,$$

which is finite if

$$2^{2s}(\rho + \epsilon) < 1,$$

or $s < \log_4 \rho$. ∎

Most of the content of [92] deals with constructing a subspace F_p which is as small as possible and contains the iterates of a suitable $\Gamma(\xi)$.

The following lemma is a partial counterpart to theorem 5.14.

LEMMA 11.12
If H satisfies the sum rules of order p, and $H(0)$ does not have an eigenvalue of the form m^{-k}, $k = p, \ldots, 2p-1$, then the transition matrix has eigenvalues m^{-k}, $k = 0, \ldots, 2p-1$.

PROOF This is shown in [92], by explicitly constructing the left eigenvectors. The details are quite lengthy. ∎

Where is the condition on the eigenvalues of $H(0)$ used in the proof of this lemma?

The sum rules of order p are conditions involving the approximation vectors $\mathbf{y}_k$, $k = 0, \ldots, p-1$, and values of H and its derivatives at the points $2\pi s/m$, $s = 0, \ldots, m-1$. If $H(0)$ does not have eigenvalues of the form m^{-k}, $k = p, \ldots, 2p-1$, we can find further vectors $\mathbf{y}_k$, $k = p, \ldots, 2p-1$ which satisfy the sum rules of corresponding order only at the point 0, but not at the other $2\pi s/m$, $s = 1, \ldots, m-1$.

In all examples considered in [92] these vectors exist, but it is conceivable that in some cases they do not. A counterexample would show that theorem 5.14 cannot be generalized to the multiwavelet case.

We note that in the scalar wavelet case $H(0)$ is a scalar; thus, the eigenvalue condition is trivially satisfied.

Based on the vectors $\mathbf{y}_k$, Jiang [92] now defines three sets of vectors: the vectors $\mathbf{y}^{(j)}$, $j = 0, \ldots, 2p-1$, are the left eigenvectors of the transition matrix to eigenvalue m^{-j}; the vectors $\mathbf{l}_s^{(j)}$ and $\mathbf{r}_s^{(j)}$, $j = 0, \ldots, p-1$, $s = 1, \ldots, r$ are not eigenvectors, and are not necessarily linearly independent, but their span is an invariant subspace. For the precise definitions of these vectors, consult the original paper.

Let F_p be the orthogonal complement of the span of the $\mathbf{y}^{(j)}$, $\mathbf{l}_s^{(j)}$, $\mathbf{r}_s^{(j)}$. Then Jiang proves

- F_p is an invariant subspace.
- $\Gamma(\xi) = (1 - \cos\xi)^{2p} I$ lies in F_p.

p should be chosen as large as possible, for a smaller space F_p.

Jiang's results give us an easy L^2-estimate, just as in the scalar case: determine the approximation order and check the eigenvalue condition for $H(0)$ from lemma 11.12; determine the eigenvalues of the transition matrix; remove the known power-of-m eigenvalues. The largest remaining eigenvalue is an upper bound on ρ, which leads to a lower bound on s.

It may not be the best possible estimate, but it is a lower bound for the actual Sobolev exponent.

Example 11.2
The DGHM multiwavelet has approximation order 2, and the transition matrix has size 28×28. Its eigenvalues include the known $1, 1/2, 1/4, 1/8$. The rest (sorted by size) are $-1/5$ (twice), $1/8$ (again), $-1/10$ (twice), $-1/20$ (four times), $1/25$, $1/50$ (twice), and 0 (twelve times).

ρ is at most 0.2. An explicit calculation of F_p shows that the eigenvectors to eigenvalue $-1/5$ are in fact not in F_p. The true ρ is no larger than $1/8$, so $s \geq 1.5$.

It can be established by other means that this ϕ is Lipschitz continuous but not differentiable, so this estimate is the best possible. □

THEOREM 11.13
Let n be the degree of $H(\xi)$.

A sufficient condition for $\phi \in L^2$ is that H satisfies the sum rules of order 1, and the transition matrix T_k satisfies condition E. Here k is the smallest integer larger than $(N-m)/(m-1)$.

The proof is identical to the scalar case.

11.3.3 Cascade Algorithm

One way of obtaining a solution to the refinement equation is to use fixed point iteration on it. If the iteration converges in L^2, this presents a practical way of approximating the function $\phi(x)$ as well as an existence proof for the solution.

DEFINITION 11.14 *Assume we are given $H(\xi)$ with $H(0) = 1$. The cascade algorithm consists of selecting a suitable starting function $\phi^{(0)}(x) \in L^2$, and then producing a sequence of functions*

$$\phi^{(n+1)}(x) = \sqrt{m} \sum_{k=k_0}^{k_1} H_k \phi^{(n)}(mx - k),$$

or equivalently

$$\hat{\phi}^{(n+1)}(\xi) = H(\xi/m)\hat{\phi}^{(n)}(\xi/m) = \Pi_n(\xi)\hat{\phi}^{(0)}(m^{-n-1}\xi).$$

THEOREM 11.15
Assume that H satisfies condition E and the sum rules of order 1 with approximation vector $\mathbf{y}_0$. Then the cascade algorithm converges for any starting function $\phi^{(0)}$ which satisfies

$$\mathbf{y}_0^* \sum_k \phi^{(0)}(k) = c \neq 0$$

if and only if the transition operator satisfies condition E.

The proof is given in [131].

11.4 Pointwise Theory

In section 6.5, we already explained how to compute point values of ϕ at the integers, and via repeated application of the refinement relation at all m-adic points. An m-adic point is a rational number of the form

$$x = \frac{k}{m^n}, \qquad n \in \mathbb{N}, \quad k \in \mathbb{Z}.$$

The following describes a more formalized way of doing this, which can then be used to obtain smoothness estimates.

To keep the notation simpler, assume that supp $\phi \subset [0, n]$. We define

$$\mathbf{\Phi}(x) = \begin{pmatrix} \phi(x) \\ \phi(x+1) \\ \vdots \\ \phi(x+n-1) \end{pmatrix}, \qquad x \in [0, 1]. \tag{11.9}$$

This is related to the $\mathbf{\Phi}$ in section 6.5, but not quite the same: $\mathbf{\Phi}(0)$ consists of the first n subvectors of the previous $\mathbf{\Phi}$, $\mathbf{\Phi}(1)$ of the last n subvectors.

The recursion relation states that

$$\phi(x+k) = \sqrt{m} \sum_\ell H_\ell \phi(mx + mk - \ell) = \sqrt{m} \sum_\ell H_{mk-\ell} \phi(mx + \ell).$$

Obviously,

$$\phi(x+k) = [\mathbf{\Phi}(x)]_k$$

(the kth subvector in $\mathbf{\Phi}(x)$), while

$$\phi(mx + \ell) = \begin{cases} [\mathbf{\Phi}(mx)]_\ell & \text{if } 0 \le x \le 1/m, \\ [\mathbf{\Phi}(mx - 1)]_{\ell+1} & \text{if } 1/m \le x \le 2/m, \\ \cdots \\ [\mathbf{\Phi}(mx - m + 1)]_{\ell+1} & \text{if } (m-1)/m \le x \le 1. \end{cases}$$

If $0 \le x \le 1/m$, then

$$[\mathbf{\Phi}(x)]_k = \sqrt{m} \sum_\ell H_{mk-\ell} [\mathbf{\Phi}(mx)]_\ell,$$

or

$$\mathbf{\Phi}(x) = T_0 \mathbf{\Phi}(mx), \tag{11.10}$$

where

$$(T_0)_{k\ell} = \sqrt{m}\, H_{mk-\ell}, \qquad 0 \le k, \ell \le n - 1.$$

If $1/m \le x \le 2/m$, then

$$[\mathbf{\Phi}(x)]_k = \sqrt{m} \sum_\ell H_{mk-\ell+1} [\mathbf{\Phi}(mx - 1)]_\ell,$$

or

$$\mathbf{\Phi}(x) = T_1 \mathbf{\Phi}(mx - 1), \tag{11.11}$$

where

$$(T_1)_{k\ell} = \sqrt{m}\, H_{mk-\ell+1}, \qquad 0 \le k, \ell \le n - 1,$$

and so on.

The matrices $T_0, \ldots, T_{m-1}$ are related to the matrix T from section 6.5. T_0 is the $n \times n$ submatrix of T at the top left. T_k is the submatrix k steps to the left, where we extend the rows of T by periodicity.

Chapter 11: Existence and Regularity

If H satisfies the sum rules of order 1, then
$$\mathbf{e}^* = (\boldsymbol{\mu}_0^*, \boldsymbol{\mu}_0^*, \ldots, \boldsymbol{\mu}_0^*)$$
is a common left eigenvector to eigenvalue 1 of all T_k.

Choose an m-adic number x. In m-adic notation, we can express x as
$$x = (0.d_1 d_2 \ldots d_k)_m, \qquad d_i \in \{0, \ldots, m-1\}.$$

Define the shift operator τ by
$$\tau x = (0.d_2 d_3 \ldots d_k)_m;$$
then
$$\tau x = \begin{cases} mx & \text{if } 0 \le x \le 1/m, \\ \ldots \\ mx - m + 1 & \text{if } (m-1)/m \le x \le 1. \end{cases}$$

Equations (11.10) and (11.11), etc. together are equivalent to
$$\boldsymbol{\Phi}(x) = T_{d_1} \boldsymbol{\Phi}(\tau x),$$
or after repeated application
$$\boldsymbol{\Phi}(x) = T_{d_1} \cdots T_{d_k} \boldsymbol{\Phi}(0). \tag{11.12}$$

Example 11.3
For the Chui–Lian multiwavelet CL2, the refinement matrices have the form
$$T_0 = \sqrt{2} \begin{pmatrix} H_0 & 0 \\ H_2 & H_1 \end{pmatrix}, \qquad T_1 = \sqrt{2} \begin{pmatrix} H_1 & H_0 \\ 0 & H_2 \end{pmatrix}.$$
The vector $\boldsymbol{\Phi}(0)$ is $(0, 0, 1, 0)^*$.

Then
$$\boldsymbol{\Phi}(1/4) = \boldsymbol{\Phi}((0.01)_2) = T_0 T_1 \boldsymbol{\Phi}(0) = \frac{1}{16} \begin{pmatrix} 4 - 2\sqrt{7} \\ 7 - 2\sqrt{7} \\ 12 + 2\sqrt{7} \\ 7 + 4\sqrt{7} \end{pmatrix}. \qquad \square$$

The same approach can be used to find point values of the multiwavelet function. We simply define matrices $S_0, S_1, \ldots$ analogous to $T_0, T_1, \ldots$
$$(S_j)_{k\ell} = \sqrt{m}\, G_{mk-\ell+j}, \qquad 0 \le k, \ell \le n-1$$
and replace the leftmost matrix in equation (11.12):
$$\boldsymbol{\Psi}(x) = S_{d_1} T_{d_2} \cdots T_{d_k} \boldsymbol{\Phi}(0).$$

This approach can also be used to find exact point values of refinable functions at any rational point, but we are interested in smoothness estimates.

DEFINITION 11.16 *The (uniform) joint spectral radius of* $T_0, \ldots, T_{m-1}$ *is defined as*

$$\rho(T_0, \ldots, T_{m-1}) = \limsup_{\ell \to \infty} \max_{d_j \in \{0, \ldots, m-1\}} \|T_{d_1} \cdots T_{d_\ell}\|^{1/\ell}.$$

THEOREM 11.17
Assume $H(\xi)$ *satisfies the sum rules of order 1, and that*

$$\rho(T_0|F_1, \ldots, T_{m-1}|F_1) = \lambda < 1.$$

where F_1 *is the orthogonal complement of the common left eigenvector* $\mathbf{e}^* = (\boldsymbol{\mu}_0^*, \ldots, \boldsymbol{\mu}_0^*)$ *of* $T_0, \ldots, T_{m-1}$.

Then the recursion relation has a unique solution ϕ *which is Hölder continuous of order* α *for any*

$$\alpha < -\log_m \lambda.$$

The proof is completely analogous to that of theorem 5.21.

As in the scalar case, generalizations of theorem 11.17 can be used to guarantee higher orders of smoothness.

THEOREM 11.18
Assume that $\phi(x)$ *has approximation order* p. *Let* F_p *be the orthogonal complement of* $\mathrm{span}\{\mathbf{e}_0, \ldots, \mathbf{e}_{p-1}\}$, *where*

$$\mathbf{e}_k = (\mathbf{e}_{k0}^*, \mathbf{e}_{k1}^*, \ldots, \mathbf{e}_{k,m-1}^*),$$

$$\mathbf{e}_{kl}^* = \int (x+j)^k \phi(x)^* \, dx = \sum_{s=0}^{k} \binom{k}{s} j^s \boldsymbol{\mu}_{k-s}^*,$$

and assume that

$$\rho(T_0|F_p, \ldots, T_{m-1}|F_p) = \lambda < 1.$$

Then $\phi \in C^\alpha$ *for any* $\alpha < -\log_m \lambda$.

This is proved in [90].

In general, a joint spectral radius can be quite hard to compute. With some effort, you can get reasonably good estimates. From the definition we get the estimate

$$\rho(T_0, \ldots, T_{m-1}) \le \max_k \|T_k\|,$$

or more generally

$$\rho(T_0, \ldots, T_{m-1}) \le \max_{d_j \in \{0, \ldots, m-1\}} \|T_{d_1} \cdots T_{d_\ell}\|^{1/\ell}$$

for any fixed ℓ. Moderately large ℓ give good bounds in practice, but the number of norms you have to examine grows like m^ℓ.

You can also get lower bounds on ρ this way, which puts upper bounds on the smoothness. For any specific choice of d_j,

$$\rho(T_0, \ldots, T_{m-1}) \geq \|T_{d_1} \cdots T_{d_\ell}\|^{1/\ell}.$$

The frequency domain methods can only give lower bounds on the smoothness estimates. The pointwise method can give both upper and lower bounds.

11.5 Smoothness and Approximation Order

We saw in the preceding sections that approximation order is important for smoothness estimates. A high approximation order does not directly guarantee smoothness, but the two tend to be correlated.

As in the scalar case, smoothness implies a certain minimum approximation order.

THEOREM 11.19
Assume that $\boldsymbol{\phi}$ is a stable refinable function vector. If $\boldsymbol{\phi}$ is p times continuously differentiable, then $\boldsymbol{\phi}$ has approximation order $p + 1$.

The proof is similar to the scalar case. It can be found in [41].

11.6 Stability

The previous sections discussed the existence of $\boldsymbol{\phi}(x)$ and its smoothness properties. In order to ensure that $\boldsymbol{\phi}$ produces an MRA we need to also verify that

- $\boldsymbol{\phi}$ has stable shifts.
- $\bigcap_k V_k = \{0\}$.
- $\overline{\bigcup_k V_k} = L^2$.

We will now give sufficient conditions for these properties.

Recall that ϕ has stable shifts if $\phi \in L^2$ and if there exist constants $0 < A \leq B$ so that for all sequences $\{\mathbf{c}_k\} \in \ell_2$,

$$A \sum_k \|\mathbf{c}_k\|^2 \leq \|\sum_k \mathbf{c}_k^* \phi(x-k)\|_2^2 \leq B \sum_k \|\mathbf{c}_k\|^2.$$

LEMMA 11.20
ϕ has stable shifts if and only if there exist constants $0 < A \leq B$ so that

$$A\|\mathbf{c}\|_2^2 \leq \sum_k \|\mathbf{c}^* \hat{\phi}(\xi + 2k\pi)\|^2 \leq B\|\mathbf{c}\|_2^2$$

for any vector $\mathbf{c}$.

This lemma is often phrased as "ϕ has stable shifts if and only if the autocorrelation matrix is strictly positive definite."

PROOF The proof proceeds as in the scalar case (lemma 5.24). Details can be found in [70] and [88]. ∎

LEMMA 11.21
If ϕ, $\tilde{\phi} \in L^2$ are biorthogonal, they have stable shifts.

The proof is identical to the scalar case (lemma 5.25).

A concept related to stability is *linear independence*. A compactly supported multiscaling function ϕ has *linearly independent shifts* if

$$\sum_j \mathbf{a}_j^* \phi(x-j) = 0 \Rightarrow \mathbf{a} = 0 \tag{11.13}$$

for all sequences $\mathbf{a}$.

It is shown in [88] and [89] that ϕ has linearly independent shifts if and only if the sequences

$$\{\hat{\phi}_j(\xi + 2\pi k)\}_{k \in \mathbb{Z}} \tag{11.14}$$

are linearly independent for all $\xi \in \mathbb{C}$. ϕ has stable shifts if and only if the sequences in equation (11.14) are linearly independent for all $\xi \in \mathbb{R}$.

Thus, linear independence implies stability.

Further stability conditions are given in [82] and [122]. Paper [93] discusses stability for the case where the refinement equation has multiple linearly independent solutions.

THEOREM 11.22
Assume $\phi \in L^2$. Then

$$\overline{\bigcup_k V_k} = L^2$$

Chapter 11: Existence and Regularity

if and only if

$$\bigcap_{k,n} m^n Z(\hat{\phi}_k)$$

is a set of measure 0, where $Z(f)$ = set of zeros of f, or equivalently if

$$\bigcup_k \operatorname{supp} \hat{\phi}_k = \mathbb{R}$$

up to a set of measure zero.

This is proved in [91].
The conditions of the theorem are satisfied if ϕ has compact support.

THEOREM 11.23
Assume $\phi \in L^2$. If

$$\sum_k |\hat{\phi}_k(\xi)| > 0$$

in some neighborhood of 0, then

$$\bigcap_n V_n = \{0\}.$$

This is shown in [69]. The conditions of this theorem are satisfied if ϕ is refinable and stable.

All the conditions for the existence of a pair of biorthogonal MRAs are satisfied if ϕ, $\tilde{\phi}$ are compactly supported, biorthogonal L^2-functions.

For a full justification of the decomposition and reconstruction, we need to show that the multiwavelet functions are also stable.

There are actually two definitions of stability that we need to consider. Stability at a single level means that there exist constants $0 < A \le B$ so that

$$A \sum_k \|\mathbf{c}_k\|^2 \le \| \sum_{kt} \mathbf{c}_k^* \psi^{(t)}(x - k)\|_2^2 \le B \sum_k \|\mathbf{c}_k\|^2.$$

This condition is automatically satisfied if the multiwavelet functions are biorthogonal. The proof is the same as that in lemma 11.21. Stability at level 0 implies stability at any other fixed level n, which is already enough to justify a decomposition and reconstruction over a finite number of levels.

Stability over all levels means

$$A \sum_{nk} \|\mathbf{c}_{nk}\|^2 \le \| \sum_{nkt} \mathbf{c}_{nk}^* m^{n/2} \psi^{(t)}(m^n x - k)\|_2^2 \le B \sum_{nk} \|\mathbf{c}_{nk}\|^2.$$

This is required if you want to decompose an L^2-function f in multiwavelet functions over all levels.

There ought to be a counterpart to theorem 5.27 for the multiwavelet case, but I have not been able to find an explicit statement of it.

A

Standard Wavelets

In all listings, p = approximation order, s = Sobolev exponent, and α = Hölder exponent. If α is not listed, $s - 1/2$ is a lower bound for α. In many cases, a common factor for all coefficients in its column is listed separately, for easier readability.

For scalar wavelets, we only list h_k and $\tilde{h}_k$. The coefficients g_k and $\tilde{g}_k$ can be found by reversing $\tilde{h}_k$ and h_k with alternating sign.

All multiwavelets listed here have $m = 2$, $r = 2$.

A.1 Scalar Orthogonal Wavelets

Daubechies Wavelets D_p

Restrictions: $p \geq 1$ integer.

Support $[0, 2p-1]$, approximation order p; values of s, α are given in the tables below (see [49] and [50].)

	h_k	h_k	h_k
p	1	2	3
$k = 0$	1	$1 + \sqrt{3}$	$1 + \sqrt{10} + \sqrt{5 + 2\sqrt{10}}$
1	1	$3 + \sqrt{3}$	$5 + \sqrt{10} + 3\sqrt{5 + 2\sqrt{10}}$
2		$3 - \sqrt{3}$	$10 - 2\sqrt{10} + 2\sqrt{5 + 2\sqrt{10}}$
3		$1 - \sqrt{3}$	$10 - 2\sqrt{10} - \sqrt{5 + 2\sqrt{10}}$
4			$5 + \sqrt{10} - 3\sqrt{5 + 2\sqrt{10}}$
5			$1 + \sqrt{10} - \sqrt{5 + 2\sqrt{10}}$
factor	$1/\sqrt{2}$	$1/(4\sqrt{2})$	$1/(16\sqrt{2})$
s	0.500	1.000	1.415
α		0.550	1.088

	h_k	h_k	h_k
p	4	5	6
$k=0$	0.23037781330890	0.16010239797419	0.11154074335011
1	0.71484657055291	0.60382926979719	0.49462389039846
2	0.63088076792986	0.72430852843777	0.75113390802110
3	−0.02798376941686	0.13842814590132	0.31525035170919
4	−0.18703481171909	−0.24229488706638	−0.22626469396544
5	0.03522629188571	−0.03224486958464	−0.12976686756726
6	0.03288301166689	0.07757149384005	0.09750160558733
7	−0.01059740178507	−0.00624149021280	0.02752286553031
8		−0.01258075199908	−0.03158203931749
9		0.00333572528547	0.00055384220116
10			0.00477725751095
11			−0.00107730108531
s	1.775	2.096	2.388
α	1.618	1.596	1.888

Coiflets

We only list the coiflets of length 6 with support $[-2,3]$ (two different coiflets) or $[-1,4]$ (another two). Coiflets on $[-3,2]$ and $[-4,1]$ can be found by reversing coefficients.

$p = 2$, $\mu_1 = \mu_2 = 0$; values of s are given in the table below (see [30], [50], and [51].)

	support $[-2,3]$		support $[-1,4]$	
	h_k	h_k	h_k	h_k
$k=-2$	$1+\sqrt{7}$	$1-\sqrt{7}$		
−1	$5-\sqrt{7}$	$5+\sqrt{7}$	$9-\sqrt{15}$	$9+\sqrt{15}$
0	$14-2\sqrt{7}$	$14+2\sqrt{7}$	$13+\sqrt{15}$	$13-\sqrt{15}$
1	$14+2\sqrt{7}$	$14-2\sqrt{7}$	$6+2\sqrt{15}$	$6-2\sqrt{15}$
2	$1+\sqrt{7}$	$1-\sqrt{7}$	$6-2\sqrt{15}$	$6+2\sqrt{15}$
3	$-3-\sqrt{7}$	$-3+\sqrt{7}$	$1-\sqrt{15}$	$1+\sqrt{15}$
4			$-3+\sqrt{15}$	$-3-\sqrt{15}$
factor	$1/(16\sqrt{2})$	$1/(16\sqrt{2})$	$1/(16\sqrt{2})$	$1/(16\sqrt{2})$
s	0.5896	1.0217	1.2321	0.0413

A.2 Scalar Biorthogonal Wavelets

Cohen(p,p̃) (Cohen–Daubechies–Feauveau)

Restrictions: p, $\tilde{p}$ integers ≥ 1, $p + \tilde{p}$ even.

ϕ is the B-spline of order p. ϕ, $\tilde{\phi}$ are symmetric about 0 (p even) or 1/2 (p odd). Approximation orders p and $\tilde{p}$ (see [40] and [50].)

There are too many of them to list here. A table of some of them is in [50, page 277]. Cohen(1,1) = Haar wavelet; Cohen(2,4) is given in example 1.6.

Daubechies(7,9)

Symmetric about 0, $p = \tilde{p} = 4$, $s = 2.1226$, $\tilde{s} = 1.4100$ (see [50].)

	h_k	$\tilde{h}_k$
$k = -4$		0.03782845550700
-3	-0.06453888262894	-0.02384946501938
-2	-0.04068941760956	-0.11062440441842
-1	0.41809227322221	0.37740285561265
0	0.78848561640566	0.85269867900940
1	0.41809227322221	0.37740285561265
2	-0.04068941760956	-0.11062440441842
3	-0.06453888262894	-0.02384946501938
4		0.03782845550700

A.3 Orthogonal Multiwavelets

BAT O1 (Lebrun–Vetterli)

Support $[0, 2]$. ϕ_2 is reflection of ϕ_1 about $x = 1$ and vice versa; wavelet functions are symmetric/antisymmetric about $x = 1$. $p = 2$, balanced of order 1, $s = 0.6406$.

There are also BAT O2 and BAT O3, which are balanced of order 2 and 3 (see [105] for a list of coefficients.)

	H_k	G_k
$k = 0$	$\begin{pmatrix} 0 & 2+\sqrt{7} \\ 0 & 2-\sqrt{7} \end{pmatrix}$	$\begin{pmatrix} 0 & -2 \\ 0 & 1 \end{pmatrix}$
1	$\begin{pmatrix} 3 & 1 \\ 1 & 3 \end{pmatrix}$	$\begin{pmatrix} 2 & 2 \\ -\sqrt{7} & \sqrt{7} \end{pmatrix}$
2	$\begin{pmatrix} 2-\sqrt{7} & 0 \\ 2+\sqrt{7} & 0 \end{pmatrix}$	$\begin{pmatrix} -2 & 0 \\ -1 & 0 \end{pmatrix}$
factor	$1/(4\sqrt{2})$	$1/4$

CL2(t) (Chui–Lian)

Restriction: $-1/\sqrt{2} \leq t < -1/2$.

Support $[0, 2]$, symmetric/antisymmetric about $x = 1$, $\alpha = -\log_2 |1/2 + 2t|$, $p = 1$.

This is a special case of JRZB(s,t,λ,μ).

Special case: CL2 = CL2$(-\sqrt{7}/4)$ is the standard Chui–Lian multiwavelet. $p = 2$, $s = 1.0545$, $\alpha = -\log_2(\sqrt{7}/4) \approx 0.59632$ (see [36] and [90].)

	H_k	G_k
$k = 0$	$\begin{pmatrix} 1 & 1 \\ 2t & 2t \end{pmatrix}$	$\begin{pmatrix} -1 & -1 \\ \mu & \mu \end{pmatrix}$
1	$\begin{pmatrix} 2 & 0 \\ 0 & 2\mu \end{pmatrix}$	$\begin{pmatrix} 2 & 0 \\ 0 & -4t \end{pmatrix}$
2	$\begin{pmatrix} 1 & -1 \\ -2t & 2t \end{pmatrix}$	$\begin{pmatrix} -1 & 1 \\ -\mu & \mu \end{pmatrix}$
factor	$1/(2\sqrt{2})$	$1/(2\sqrt{2})$

where

$$\mu = \sqrt{2 - 4t^2}.$$

CL3 (Chui–Lian)

Support $[0, 3]$, symmetric/antisymmetric about $x = 3/2$, $p = 3$, $s = 1.4408$ (see [36].)

	H_k	G_k
$k = 0$	$\begin{pmatrix} 10 - 3\sqrt{10} & 5\sqrt{6} - 2\sqrt{15} \\ 5\sqrt{6} - 3\sqrt{15} & 5 - 3\sqrt{10} \end{pmatrix}$	$\begin{pmatrix} 5\sqrt{6} - 2\sqrt{15} & -10 + 3\sqrt{10} \\ -5 + 3\sqrt{10} & 5\sqrt{6} - 3\sqrt{15} \end{pmatrix}$
1	$\begin{pmatrix} 30 + 3\sqrt{10} & 5\sqrt{6} - 2\sqrt{15} \\ -5\sqrt{6} - 7\sqrt{15} & 15 - 3\sqrt{10} \end{pmatrix}$	$\begin{pmatrix} -5\sqrt{6} + 2\sqrt{15} & 30 + 3\sqrt{10} \\ 15 - 3\sqrt{10} & 5\sqrt{6} + 7\sqrt{15} \end{pmatrix}$
2	$\begin{pmatrix} 30 + 3\sqrt{10} & -5\sqrt{6} + 2\sqrt{15} \\ 5\sqrt{6} + 7\sqrt{15} & 15 - 3\sqrt{10} \end{pmatrix}$	$\begin{pmatrix} -5\sqrt{6} + 2\sqrt{15} & -30 - 3\sqrt{10} \\ -15 + 3\sqrt{10} & 5\sqrt{6} + 7\sqrt{15} \end{pmatrix}$
3	$\begin{pmatrix} 10 - 3\sqrt{10} & -5\sqrt{6} + 2\sqrt{15} \\ -5\sqrt{6} + 3\sqrt{15} & 5 - 3\sqrt{10} \end{pmatrix}$	$\begin{pmatrix} 5\sqrt{6} - 2\sqrt{15} & 10 - 3\sqrt{10} \\ 5 - 3\sqrt{10} & 5\sqrt{6} - 3\sqrt{15} \end{pmatrix}$
factor	$1/(40\sqrt{2})$	$1/(40\sqrt{2})$

Balanced Daubechies D_p

Restriction: $p \geq 1$ integer.

Support $[0, 2p]$, balanced of order p, same smoothness properties as scalar Daubechies wavelets D_p.

These multiwavelets use the same coefficients as the scalar Daubechies wavelets, sorted into two rows. ϕ_1 is the Daubechies ϕ compressed by a factor of 2. ϕ_2 is ϕ_1 shifted right by $1/2$. The same holds for ψ_1, ψ_2 (see [105].)

Appendix A: Standard Wavelets

	H_k	G_k
$k=0$	$\begin{pmatrix} h_0 & h_1 \\ 0 & 0 \end{pmatrix}$	$\begin{pmatrix} g_0 & g_1 \\ 0 & 0 \end{pmatrix}$
1	$\begin{pmatrix} h_2 & h_3 \\ h_0 & h_1 \end{pmatrix}$	$\begin{pmatrix} g_2 & g_3 \\ g_0 & g_1 \end{pmatrix}$
2	$\begin{pmatrix} h_4 & h_5 \\ h_2 & h_3 \end{pmatrix}$	$\begin{pmatrix} g_4 & g_5 \\ g_2 & g_3 \end{pmatrix}$
$\vdots$		
p	$\begin{pmatrix} 0 & 0 \\ h_{2p-2} & h_{2p-1} \end{pmatrix}$	$\begin{pmatrix} 0 & 0 \\ g_{2p-2} & g_{2p-1} \end{pmatrix}$

DGHM (Donovan–Geronimo–Hardin–Massopust)

Support $[0,2]$, $p=2$, $\alpha=1$, $s=1.5$. ϕ_1 is symmetric about $x=1/2$, ϕ_2 is symmetric about $x=1$ (see [59].)

	H_k	G_k
$k=0$	$\begin{pmatrix} 12 & 16\sqrt{2} \\ -\sqrt{2} & -6 \end{pmatrix}$	$\begin{pmatrix} -\sqrt{2} & -6 \\ 2 & 6\sqrt{2} \end{pmatrix}$
1	$\begin{pmatrix} 12 & 0 \\ 9\sqrt{2} & 20 \end{pmatrix}$	$\begin{pmatrix} 9\sqrt{2} & -20 \\ -18 & 0 \end{pmatrix}$
2	$\begin{pmatrix} 0 & 0 \\ 9\sqrt{2} & -6 \end{pmatrix}$	$\begin{pmatrix} 9\sqrt{2} & -6 \\ 18 & -6\sqrt{2} \end{pmatrix}$
3	$\begin{pmatrix} 0 & 0 \\ -\sqrt{2} & 0 \end{pmatrix}$	$\begin{pmatrix} -\sqrt{2} & 0 \\ -2 & 0 \end{pmatrix}$
factor	$1/(20\sqrt{2})$	$1/(20\sqrt{2})$

STT (Shen–Tan–Tam)

Support $[0,3]$, symmetric/antisymmetric about $x=3/2$, $p=1$, $s=0.9919$ (see [130].)

	H_k	G_k
$k=0$	$\begin{pmatrix} 1 & 4+\sqrt{15} \\ 1 & -4-\sqrt{15} \end{pmatrix}$	$\begin{pmatrix} -4-\sqrt{15} & 1 \\ -4-\sqrt{15} & -1 \end{pmatrix}$
1	$\begin{pmatrix} 31+8\sqrt{15} & 4+\sqrt{15} \\ -31-8\sqrt{15} & 4+\sqrt{15} \end{pmatrix}$	$\begin{pmatrix} 4+\sqrt{15} & -31-8\sqrt{15} \\ -4-\sqrt{15} & -31-8\sqrt{15} \end{pmatrix}$
2	$\begin{pmatrix} 31+8\sqrt{15} & -4-\sqrt{15} \\ 31+8\sqrt{15} & 4+\sqrt{15} \end{pmatrix}$	$\begin{pmatrix} 4+\sqrt{15} & 31+8\sqrt{15} \\ 4+\sqrt{15} & -31-8\sqrt{15} \end{pmatrix}$
3	$\begin{pmatrix} 1 & -4-\sqrt{15} \\ -1 & -4-\sqrt{15} \end{pmatrix}$	$\begin{pmatrix} -4-\sqrt{15} & -1 \\ 4+\sqrt{15} & -1 \end{pmatrix}$
factor	$1/(8(4+\sqrt{15}))$	$1/(8(4+\sqrt{15}))$

A.4 Biorthogonal Multiwavelets

HM(s) (Hardin–Marasovich)
   Restriction: $-1 < s < 1/7$.
   Support $[-1, 1]$; ϕ_1, $\tilde{\phi}_1$ are symmetric about $x = 0$; ϕ_2, $\tilde{\phi}_2$ are symmetric about $x = 1/2$; $p = \tilde{p} = 2$ (see [75].)

	H_k	G_k
$k = -2$	$\begin{pmatrix} 0 & a \\ 0 & 0 \end{pmatrix}$	$\begin{pmatrix} 0 & a \\ 0 & \sqrt{2}a \end{pmatrix}$
-1	$\begin{pmatrix} b & c \\ 0 & 0 \end{pmatrix}$	$\begin{pmatrix} b & c \\ \sqrt{2}b & \sqrt{2}c \end{pmatrix}$
0	$\begin{pmatrix} 1 & c \\ 0 & d \end{pmatrix}$	$\begin{pmatrix} -1 & c \\ 0 & -\sqrt{2}c \end{pmatrix}$
1	$\begin{pmatrix} b & a \\ e & d \end{pmatrix}$	$\begin{pmatrix} b & a \\ -\sqrt{2}b & -\sqrt{2}a \end{pmatrix}$
factor	$1/\sqrt{2}$	$1/\sqrt{2}$

where

$$\tilde{s} = (1+2s)/(-2+5s) \Leftrightarrow s = (1+2\tilde{s})/(-2+5\tilde{s})$$
$$\alpha = 3(1-s)(1-s\tilde{s})/(4-s-\tilde{s}-2s\tilde{s})$$
$$\gamma = \sqrt{6(4-s-\tilde{s}-2s\tilde{s})/(7-4s-4\tilde{s}+s\tilde{s})}$$
$$\delta = \sqrt{12(-1+\tilde{s})(-1+s)(-1+s\tilde{s})/(-4+s+\tilde{s}+2s\tilde{s})}$$
$$a = \alpha\gamma(1-2\alpha-2s)/(2\delta)$$
$$b = 1/2 - \alpha$$
$$c = \alpha\gamma(3-2\alpha-2s)/(2\delta)$$
$$d = \alpha + s$$
$$e = \delta/\gamma$$

   The dual functions have the same form. Exchange tilde and nontilde in all formulas.
   Special case: for $s = 0$, $V_0 =$ continuous piecewise linear splines with half-integer knots.
   Special case: for $s = 1/4$, $V_0 =$ continuous, piecewise quadratic splines with integer knots.

JRZB(s,t,λ,μ) (Jia–Riemenschneider–Zhou biorthogonal)
   Restriction: $|2\lambda + \mu| < 2$.
   Support $[0, 2]$, symmetric/antisymmetric about $x = 1$, $p = 1$, $\alpha = 2$ if $|st + 1/4| \leq 1/8$, $\alpha = -\log_2 |2st + 1/2|$ if $1/8 < |st + 1/4| < 1/2$ (see [90].)
   Special case: $p = 3$ if $t \neq 0$, $\mu = 1/2$, $\lambda = 2st + 1/4$.

Appendix A: Standard Wavelets

Special case: $p = 4$ if $\lambda = -1/8$, $\mu = 1/2$, $st = -3/16$. For $s = 3/2$, $t = -1/8$ we get the cubic Hermite multiwavelet.

Special case: $s = 1$, $\lambda = t$, $\mu = \sqrt{2-4t^2}$ is CL2(t).

	H_k
$k = 0$	$\begin{pmatrix} 1 & s \\ 2t & 2\lambda \end{pmatrix}$
1	$\begin{pmatrix} 2 & 0 \\ 0 & 2\mu \end{pmatrix}$
2	$\begin{pmatrix} 1 & -s \\ -2t & 2\lambda \end{pmatrix}$
factor	$1/(2\sqrt{2})$

I could not find multiwavelet or dual multiscaling function coefficients for this multiscaling function published anywhere.

HC (Hermite cubic)

This is a special case of JRZB(s,t,λ,μ).

There are many completions. The one listed here is the smoothest symmetric completion with support length 4 (see [80].)

Support of ϕ is $[-1, 1]$, support of $\tilde{\phi}$ is $[-2, 2]$; all functions are symmetric/antisymmetric about $x = 0$; $p = 4$, $\tilde{p} = 2$, $\alpha = 2$, $s = 2.5$, $\tilde{s} = 0.8279$.

	H_k	G_k	$\tilde{H}_k$	$\tilde{G}_k$
$k = -2$			$\begin{pmatrix} -2190 & -1540 \\ 13914 & 9687 \end{pmatrix}$	
-1	$\begin{pmatrix} 4 & 6 \\ -1 & -1 \end{pmatrix}$	$\begin{pmatrix} 5427 & 567 \\ -1900 & -120 \end{pmatrix}$	$\begin{pmatrix} 9720 & 3560 \\ -60588 & -21840 \end{pmatrix}$	
0	$\begin{pmatrix} 8 & 0 \\ 0 & 4 \end{pmatrix}$	$\begin{pmatrix} -19440 & -60588 \\ 7120 & 21840 \end{pmatrix}$	$\begin{pmatrix} 23820 & 0 \\ 0 & 36546 \end{pmatrix}$	$\begin{pmatrix} -2 & -1 \\ 3 & 1 \end{pmatrix}$
1	$\begin{pmatrix} 4 & -6 \\ 1 & -1 \end{pmatrix}$	$\begin{pmatrix} 28026 & 0 \\ 0 & 56160 \end{pmatrix}$	$\begin{pmatrix} 9720 & -3560 \\ 60588 & -21840 \end{pmatrix}$	$\begin{pmatrix} 4 & 0 \\ 0 & 4 \end{pmatrix}$
2		$\begin{pmatrix} -19440 & 60588 \\ -7120 & 21840 \end{pmatrix}$	$\begin{pmatrix} -2190 & 1540 \\ -13914 & 9687 \end{pmatrix}$	$\begin{pmatrix} -2 & 1 \\ -3 & 1 \end{pmatrix}$
3		$\begin{pmatrix} 5427 & -567 \\ 1900 & -120 \end{pmatrix}$		
factor	$1/(8\sqrt{2})$	$1/(19440\sqrt{2})$	$1/(19440\sqrt{2})$	$1/(8\sqrt{2})$

B

Mathematical Background

B.1 Notational Conventions

This book uses the following notational convention: scalars are usually written as lowercase letters (Roman or Greek). Vectors are written as boldface lowercase letters. Matrices are written as uppercase letters.

Vectors are always column vectors; row vectors are written as transposes of column vectors.

The notation a^* denotes the complex conjugate of a. For vectors and matrices, $\mathbf{a}^*$ or A^* is the complex conjugate transpose of $\mathbf{a}$ or A.

B.2 Derivatives

The derivative operator is usually written as D. We will often need the following formula (repeated product rule):

LEMMA B.1 Leibniz Rule
If $f(x)$, $g(x)$ are n times continuously differentiable, then

$$D^n [f(x)g(x)] = \sum_{k=0}^{n} \binom{n}{k} D^k f(x) D^{n-k} g(x).$$

B.3 Functions and Sequences

All functions in this book are considered to be complex-valued functions of a real argument. This way, all the formulas will contain complex conjugates in the right places. In most applications the functions are in fact real-valued, so the complex conjugate transpose can be ignored in the scalar case, or read as a real transpose in the vector or matrix case.

The natural setting for wavelet theory is the space $L^2 = L^2(\mathbb{R})$ of square-integrable functions. These are functions whose L^2-norm

$$\|f\|_2 = \left(\int_{-\infty}^{\infty} |f(x)|^2 \, dx \right)^{1/2}$$

is finite. The inner product on L^2 is

$$\langle f, g \rangle = \int_{-\infty}^{\infty} f(x) g(x)^* \, dx.$$

For vectors $\mathbf{f}, \mathbf{g}$, the inner product is defined analogously as

$$\langle \mathbf{f}, \mathbf{g} \rangle = \int_{-\infty}^{\infty} \mathbf{f}(x) \mathbf{g}(x)^* \, dx.$$

This inner product is a matrix

$$\langle \mathbf{f}, \mathbf{g} \rangle = \begin{pmatrix} \langle f_1, g_1 \rangle & \cdots & \langle f_1, g_n \rangle \\ \vdots & & \vdots \\ \langle f_n, g_1 \rangle & \cdots & \langle f_n, g_n \rangle \end{pmatrix}.$$

We also use the space $L^1 = L^1(\mathbb{R})$ of integrable functions. These are functions with finite L^1-norm:

$$\|f\|_1 = \int_{-\infty}^{\infty} |f(x)| \, dx < \infty.$$

The function $\chi_{[a,b]}$ is the characteristic function of the interval $[a, b]$. It is defined by

$$\chi_{[a,b]}(x) = \begin{cases} 1 & \text{if } x \in [a, b], \\ 0 & \text{otherwise.} \end{cases}$$

The space ℓ^2 consists of doubly infinite sequences of complex numbers $\{c_k\}$, $k \in \mathbb{Z}$, with

$$\|\{c_k\}\|_2 = \left(\sum_k |c_k|^2 \right)^{1/2} < \infty.$$

The space $(\ell^2)^r$ consists of doubly infinite sequences of r-vectors $\{\mathbf{c}_k\}$, with

$$\|\{\mathbf{c}_k\}\|_2 = \left(\sum_k \|\mathbf{c}_k\|^2 \right)^{1/2} = \left(\sum_{k,j} |c_{kj}|^2 \right)^{1/2} < \infty.$$

The *Kronecker delta* sequence δ_k is defined by

$$\delta_{k\ell} = \begin{cases} 1 & \text{if } k = \ell, \\ 0 & \text{otherwise.} \end{cases}$$

B.4 Fourier Transform

The Fourier transform is defined by

$$\hat{f}(\xi) = \frac{1}{\sqrt{2\pi}} \int_{-\infty}^{\infty} f(x) e^{-ix\xi} \, dx.$$

The Fourier transform of an L^1-function is a continuous function which goes to zero at infinity. The Fourier transform of an L^2-function is another L^2-function.

Some of the standard properties of the Fourier transform are listed here for easier reference. Different authors use slightly different definitions, so these formulas may differ by factors of $\sqrt{2\pi}$ or in other minor ways from formulas in other books.

Parseval's formula:

$$\langle f, g \rangle = \langle \hat{f}, \hat{g} \rangle.$$

Convolution and correlation: If h is the convolution of f and g

$$h(x) = \int f(y) g(x-y) \, dy,$$

then

$$\hat{h}(\xi) = \sqrt{2\pi} \hat{f}(\xi) \hat{g}(\xi).$$

We mostly use L^2-inner products, where the second term carries a complex conjugate transpose, so we need a slightly different statement: if h is the correlation of f and g

$$h(x) = \langle f(y), g(y-x) \rangle = \int f(y) g(y-x)^* \, dy,$$

then

$$\hat{h}(\xi) = \sqrt{2\pi} \hat{f}(\xi) \hat{g}(\xi)^*.$$

Differentiation:

$$(Df)\hat{}(\xi) = i\xi \hat{f}(\xi).$$

Multiplication by x:

$$(xf(x))\hat{}(\xi) = iD\hat{f}(\xi).$$

Translation:

$$(f(x-a))\hat{}(\xi) = e^{-i\xi} \hat{f}(\xi).$$

Poisson summation formula:

$$\sum_k f(hk) = \frac{\sqrt{2\pi}}{h} \sum_k \hat{f}(\frac{2\pi}{h} k).$$

B.5 Laurent Polynomials

A Laurent polynomial in a formal variable z has the form

$$f(z) = \sum_{k=k_0}^{k_1} f_k z^k. \tag{B.1}$$

Laurent polynomials differ from regular polynomials in that negative powers are allowed. In all important aspects they behave like regular polynomials.

The degree of the Laurent polynomial (B.1) is $k_1 - k_0$.

Vector or matrix Laurent polynomials are Laurent polynomials whose coefficients are vectors or matrices

$$\mathbf{f}(z) = \sum_{k=k_0}^{k_1} \mathbf{f}_k z^k,$$

$$F(z) = \sum_{k=k_0}^{k_1} F_k z^k.$$

Equivalently, we could consider them to be vectors or matrices with polynomial entries.

B.6 Trigonometric Polynomials

Trigonometric polynomials play a large role in wavelet theory. A trigonometric polynomial has the form

$$f(\xi) = \sum_{k=k_0}^{k_1} f_k e^{-ik\xi} \tag{B.2}$$

where k_0, k_1 may be positive or negative. All trigonometric polynomials are 2π-periodic and infinitely often differentiable. The degree of the trigonometric polynomial (B.2) is $k_1 - k_0$.

We often use the z-notation

$$f(z) = \sum_{k=k_0}^{k_1} f_k z^k, \quad z = e^{-i\xi},$$

Appendix B: Mathematical Background

which is easier to work with in many situations. In z-notation, the complex conjugate is

$$f(z)^* = \sum_{k=k_0}^{k_1} f_k^* z^{-k}.$$

There is a one-to-one correspondence between trigonometric polynomials and Laurent polynomials. We will freely switch back and forth between the two notations.

Trigonometric vector or matrix polynomials are trigonometric polynomials whose coefficients are vectors or matrices

$$\mathbf{f}(\xi) = \sum_k \mathbf{f}_k e^{-ik\xi},$$

$$F(\xi) = \sum_k F_k e^{-ik\xi}.$$

Equivalently, we could consider them to be vectors or matrices with trigonometric polynomial entries.

By using the z-notation, it is easy to see that many properties of regular polynomials carry over to trigonometric polynomials. Specifically,

- If $f(\xi)$ is nonzero for all ξ, then $1/f(\xi)$ is well-defined. $1/f(\xi)$ is a trigonometric polynomial if and only if $f(\xi)$ is a monomial; that is,

$$f(\xi) = f_n e^{-in\xi}$$

for some n, with $f_n \neq 0$.

- f has a zero of order p at a point ξ_0 if and only if it contains a factor of $(e^{-i\xi_0} - e^{-i\xi})^p$. In particular, f has a zero of order p at $\xi = 0$ if and only if it can be written as

$$f(\xi) = \left(\frac{1 - e^{-i\xi}}{2}\right)^p f_0(\xi)$$

for some trigonometric polynomial f_0. The factor 2 is introduced to ensure that $f(0) = f_0(0)$. Likewise, f has a zero of order p at $\xi = \pi$ if and only if it can be written as

$$f(\xi) = \left(\frac{1 + e^{-i\xi}}{2}\right)^p f_0(\xi).$$

B.7 Linear Algebra

DEFINITION B.2 The Frobenius norm *of a matrix A is defined as*

$$\|A\|_F^2 = \sum_{jk} |a_{jk}|^2.$$

All vector and matrix norms in finite dimensions are equivalent.

DEFINITION B.3 *A* Householder matrix *is of the form*

$$H = I - 2\mathbf{u}\mathbf{u}^*$$

for some unit vector $\mathbf{u}$.

It satisfies $H = H^* = H^{-1}$. For any vector $\mathbf{a}$ it is possible to construct a Householder matrix so that

$$H\mathbf{a} = \begin{pmatrix} \pm\|a\| \\ 0 \\ \vdots \\ 0 \end{pmatrix}.$$

DEFINITION B.4 *The* vectorization *of an $m \times n$ matrix A is defined as the vector of length mn*

$$\mathrm{vec}(A) = \begin{pmatrix} a_{11} \\ \vdots \\ a_{m1} \\ a_{21} \\ \vdots \\ a_{m2} \\ a_{31} \\ \vdots \\ a_{mn} \end{pmatrix}.$$

That is, $\mathrm{vec}(A)$ consists of the columns of A stacked on top of one another.

DEFINITION B.5 *The* Kronecker product *of two matrices A, B is de-*

fined as

$$A \otimes B = \begin{pmatrix} a_{11}B & a_{12}B & \cdots & a_{1n}B \\ a_{21}B & a_{22}B & \cdots & a_{2n}B \\ \vdots & & & \vdots \\ a_{m1}B & a_{m2}B & \cdots & a_{mn}B \end{pmatrix}.$$

LEMMA B.6
For any matrices A, B, C

$$vec(ABC) = (C^T \otimes A)vec(B).$$

DEFINITION B.7 A matrix A is positive semidefinite if

$$\mathbf{x}^* A \mathbf{x} \geq 0 \quad \text{for all } \mathbf{x}. \tag{B.3}$$

We use the notation $A \geq 0$ for this. For two matrices A, B, $A \geq B$ means $A - B \geq 0$.

If $A(\xi)$ is any matrix trigonometric polynomial, then $A(\xi)A(\xi)^*$ is positive semidefinite for all ξ.

DEFINITION B.8 The trace of a square matrix A is the sum of its diagonal terms:

$$trace(A) = \sum_k a_{kk}.$$

LEMMA B.9
If $A \leq B$, then

$$trace(A) \leq trace(B)$$

and

$$CAC^* \leq CBC^*$$

for any matrix C.
If $A(\xi) \leq B(\xi)$ for all ξ, then

$$\int A(\xi)\,d\xi \leq \int B(\xi)\,d\xi.$$

LEMMA B.10
Let E_n be the space of trigonometric matrix polynomials of the form

$$A(\xi) = \sum_{k=-n}^{n} A_k e^{-ik\xi}, \quad A_k \text{ real.}$$

Then there exists a constant c so that

$$\int_{-\pi}^{\pi} trace(A(\xi))\, d\xi \leq c\|A\|$$

C
Computer Resources

C.1 Wavelet Internet Resources

The *Wavelet Digest* is a free electronic newsletter for the wavelet community. It was originally started by Wim Sweldens, and is now managed by Michael Unser. You can find back issues and subscription information at:

www.wavelet.org

There are many web sites devoted to wavelets. Some of them come and go, some are more stable. The three sites listed here are meta-sites containing many links to other wavelet-related sites with tutorials, bibliographies, software and more, and they have all been around for many years.

- Amara Graps maintains a wavelet site at:

  www.amara.com/current/wavelet.html

- The *Wavelet Net Care* site at Washington University, Saint Louis, can be found at at:

  www.math.wustl.edu/wavelet

- Andreas Uhl is in charge of the wavelet site in Salzburg, Austria, at:

  www.cosy.sbg.ac.at/~uhl/wav.html

I am not aware of any web site specifically for multiwavelets. The listed sites all include multiwavelet material.

C.2 Wavelet Software

There are many wavelet toolboxes available. Most of them are for Matlab or are stand-alone programs, but there are tools for Mathematica, MathCAD and other systems as well.

Free packages

- WaveLab (for Matlab) was developed at Stanford University; you can find it at:

  www-stat.stanford.edu/~wavelab

- LastWave (written in C) was developed by Emmanuel Bacry; you can find it at:

  www.cmap.polytechnique.fr/~bacry/LastWave/index.html

- The Rice Wavelet Toolbox (for Matlab, with some parts in C) was developed at Rice University; it is available at:

  www-dsp.rice.edu/software/rwt.shtml

- LiftPack (written in C) was developed by Gabriel Fernández, Senthil Periaswamy and Wim Sweldens; it is available at:

  www.cs.dartmouth.edu/~sp/liftpack/lift.html

- WAILI (Wavelets with Integer Lifting) (written in C++) was developed by Geert Uytterhoeven, Filip Van Wulpen, Maarten Jansen in Leuven, Belgium; it is available at:

  www.cs.kuleuven.ac.be/~wavelets

- Angela Kunoth wrote a program package to compute integrals of refinable functions. The programs are available at:

  www.igpm.rwth-aachen.de/kunoth/prog/bw

  The documentation is available at:

  ftp://elc2.igpm.rwth-aachen.de/pub/kunoth/papers/inn.ps.Z

- The AWFD software package (Adaptivity, Wavelets and Finite Differences) (Matlab and C++) for wavelet solution of partial differential and integral equations was developed at the University of Bonn; it is available at:

  wissrech.iam.uni-bonn.de/research/projects/AWFD/index.html

Appendix C: Computer Resources 257

Commercial packages

- MathWorks it the maker of Matlab, so the MathWorks wavelet toolbox is the "official" Matlab wavelet toolbox. Their home page is:

  www.mathworks.com

- WavBox (for Matlab) was developed by Carl Taswell; you can find information at:

  www.wavbox.com.

  Early versions of this software were free. There may still be some copies archived somewhere.

- Wavelet Packet Lab (stand-alone for Microsoft Windows machines) was written by Victor Wickerhauser at Washington University, Saint Louis; you can find information at:

  www.ibuki-trading-post.com/dir_akp/akp_wavpac.html

  Free earlier versions for NeXT machines can be found at:

  www.math.wustl.edu/~victor/software/WPLab/index.html

None of these programs can handle multiwavelets, as far as I know.
A longer list of wavelet software can be found at:

  www.amara.com/current/wavesoft.html

C.3 Multiwavelet Software

This list is very short:

- MWMP (the Multiwavelet Matlab Package) was written by Vasily Strela; you can find it at:

  www.mcs.drexel.edu/~vstrela/MWMP

- My own set of Matlab routines will be available via my personal home page at:

  http://www.math.iastate.edu/keinert

and via the CRC Press download page at:

> www.crcpress.com/e_products/downloads/default.asp

My routines include a new Matlab data type `@mpoly` which represents Laurent matrix polynomials, plus routines for multiwavelet transforms, plotting multiwavelets, determining their properties, performing TSTs, lifting steps, and so on. Documentation is provided with the package.

This is a work in progress, so new features will probably be added periodically. Send bug reports or other suggestions to:

> `keinert@iastate.edu`.

An errata listing for this book will be available at the same two locations.

References

[1] B. ALPERT, G. BEYLKIN, D. GINES, AND L. VOZOVOI, *Adaptive solution of partial differential equations in multiwavelet bases*, J. Comput. Phys., 182 (2002), pp. 149–190.

[2] B. K. ALPERT, *Sparse representation of smooth linear operators*, Tech. Report YALEU/DCS/RR-814, Yale University, New Haven, CT, 1990.

[3] B. K. ALPERT, *A class of bases in L^2 for the sparse representation of integral operators*, SIAM J. Math. Anal., 24 (1993), pp. 246–262.

[4] B. K. ALPERT, G. BEYLKIN, R. COIFMAN, AND V. ROKHLIN, *Wavelets for the fast solution of second-kind integral equations*, Tech. Report YALEU/DCS/RR-837, Yale University, New Haven, CT, 1990.

[5] U. AMATO AND D. T. VUZA, *A Mathematica algorithm for the computation of the coefficients for the finite interval wavelet transform*, Rev. Roumaine Math. Pures Appl., 44 (1999), pp. 707–736.

[6] L. ANDERSSON, N. HALL, B. JAWERTH, AND G. PETERS, *Wavelets on closed subsets of the real line*, in Recent advances in wavelet analysis, Academic Press, Boston, MA, 1994, pp. 1–61.

[7] F. ARÀNDIGA, V. F. CANDELA, AND R. DONAT, *Fast multiresolution algorithms for solving linear equations: a comparative study*, SIAM J. Sci. Comput., 16 (1995), pp. 581–600.

[8] K. ATTAKITMONGCOL, D. P. HARDIN, AND D. M. WILKES, *Multiwavelet prefilters II: optimal orthogonal prefilters*. IEEE Trans. Image Proc., in press.

[9] A. AVERBUCH, E. BRAVERMAN, AND M. ISRAELI, *Parallel adaptive solution of a Poisson equation with multiwavelets*, SIAM J. Sci. Comput., 22 (2000), pp. 1053–1086.

[10] A. AVERBUCH, M. ISRAELI, AND L. VOZOVOI, *Solution of time-dependent diffusion equations with variable coefficients using multiwavelets*, J. Comput. Phys., 150 (1999), pp. 394–424.

[11] A. Z. AVERBUCH AND V. A. ZHELUDEV, *Construction of biorthogonal discrete wavelet transforms using interpolatory splines*, Appl. Comput. Harmon. Anal., 12 (2002), pp. 25–56.

[12] ——, *Lifting scheme for biorthogonal multiwavelets originated from hermite splines*, IEEE Trans. Signal Process., 50 (2002), pp. 487–500.

[13] S. BACCHELLI, M. COTRONEI, AND D. LAZZARO, *An algebraic construction of k-balanced multiwavelets via the lifting scheme*, Numer. Algorithms, 23 (2000), pp. 329–356.

[14] S. BACCHELLI, M. COTRONEI, AND T. SAUER, *Multifilters with and without prefilters*, BIT, 42 (2002), pp. 231–261.

[15] A. BARINKA, T. BARSCH, P. CHARTON, A. COHEN, S. DAHLKE, W. DAHMEN, AND K. URBAN, *Adaptive wavelet schemes for elliptic problems—implementation and numerical experiments*, SIAM J. Sci. Comput., 23 (2001), pp. 910–939.

[16] A. BARINKA, T. BARSCH, S. DAHLKE, AND M. KONIK, *Some remarks on quadrature formulas for refinable functions and wavelets*, ZAMM Z. Angew. Math. Mech., 81 (2001), pp. 839–855.

[17] A. BARINKA, T. BARSCH, S. DAHLKE, M. MOMMER, AND M. KONIK, *Quadrature formulas for refinable functions and wavelets II: error analysis*, J. Comput. Anal. Appl., 4 (2002), pp. 339–361.

[18] F. BASTIN AND C. BOIGELOT, *Biorthogonal wavelets in $H^m(\mathbf{R})$*, J. Fourier Anal. Appl., 4 (1998), pp. 749–768.

[19] L. BERG AND G. PLONKA, *Some notes on two-scale difference equations*, in Functional equations and inequalities, Kluwer, Dordrecht, 2000, pp. 7–29.

[20] S. BERTOLUZZA, C. CANUTO, AND K. URBAN, *On the adaptive computation of integrals of wavelets*, Appl. Numer. Math., 34 (2000), pp. 13–38.

[21] G. BEYLKIN, *On the representation of operators in bases of compactly supported wavelets*, SIAM J. Numer. Anal., 29 (1992), pp. 1716–1740.

[22] ——, *On the fast algorithm for multiplication of functions in the wavelet bases*, in Progress in wavelet analysis and applications (Toulouse, 1992), Frontières, Gif, 1993, pp. 53–61.

[23] ——, *On wavelet-based algorithms for solving differential equations*, in Wavelets: mathematics and applications, CRC, Boca Raton, FL, 1994, pp. 449–466.

[24] G. BEYLKIN, R. COIFMAN, AND V. ROKHLIN, *Fast wavelet transforms and numerical algorithms I*, Comm. Pure Appl. Math., 44 (1991), pp. 141–183.

[25] ——, *Wavelets in numerical analysis*, in Wavelets and their applications, Jones & Bartlett, Boston, MA, 1992, pp. 181–210.

[26] J. H. BRAMBLE, J. E. PASCIAK, AND J. XU, *Parallel multilevel preconditioners*, Math. Comp., 55 (1990), pp. 1–22.

[27] C. BRISLAWN, *Classification of nonexpansive symmetric extension transforms for multirate filter banks*, Appl. Comput. Harmon. Anal., 3 (1996), pp. 337–357.

[28] C. M. BRISLAWN, *Fingerprints go digital*, Notices Am. Math. Soc., 42 (1995), pp. 1278–1283.

[29] T. D. BUI AND G. CHEN, *Translation-invariant denoising using multiwavelets*, IEEE Trans. Signal Process., 46 (1998), pp. 3414–3420.

[30] C. S. BURRUS, R. A. GOPINATH, AND H. GUO, *Introduction to wavelets and wavelet transforms: a primer*, Prentice Hall, New York, 1998.

[31] C. S. BURRUS AND J. E. ODEGARD, *Coiflet systems and zero moments*, IEEE Trans. Signal Process., 46 (1998), pp. 761–766.

[32] C. CANUTO, A. TABACCO, AND K. URBAN, *The wavelet element method I: construction and analysis*, Appl. Comput. Harmon. Anal., 6 (1999), pp. 1–52.

[33] ———, *The wavelet element method II: realization and additional features in 2D and 3D*, Appl. Comput. Harmon. Anal., 8 (2000), pp. 123–165.

[34] Z. CHEN, C. A. MICCHELLI, AND Y. XU, *The Petrov–Galerkin method for second kind integral equations II: multiwavelet schemes*, Adv. Comput. Math., 7 (1997), pp. 199–233.

[35] Z. CHEN AND Y. XU, *The Petrov–Galerkin and iterated Petrov–Galerkin methods for second-kind integral equations*, SIAM J. Numer. Anal., 35 (1998), pp. 406–434.

[36] C. K. CHUI AND J.-A. LIAN, *A study of orthonormal multi-wavelets*, Appl. Numer. Math., 20 (1996), pp. 273–298.

[37] A. COHEN, *Wavelet methods in numerical analysis*, in Handbook of numerical analysis, Vol. VII, Handb. Numer. Anal., VII, North-Holland, Amsterdam, 2000, pp. 417–711.

[38] ———, *Numerical analysis of wavelet methods*, Vol. 32 of Studies in mathematics and its applications, Elsevier, 2003.

[39] A. COHEN AND I. DAUBECHIES, *A stability criterion for biorthogonal wavelet bases and their related subband coding scheme*, Duke Math. J., 68 (1992), pp. 313–335.

[40] A. COHEN, I. DAUBECHIES, AND J.-C. FEAUVEAU, *Biorthogonal bases of compactly supported wavelets*, Comm. Pure Appl. Math., 45 (1992), pp. 485–560.

[41] A. COHEN, I. DAUBECHIES, AND G. PLONKA, *Regularity of refinable function vectors*, J. Fourier Anal. Appl., 3 (1997), pp. 295–324.

[42] A. COHEN, I. DAUBECHIES, AND P. VIAL, *Wavelets on the interval and fast wavelet transforms*, Appl. Comput. Harmon. Anal., 1 (1993), pp. 54–81.

[43] D. COLELLA AND C. HEIL, *Characterizations of scaling functions: continuous solutions*, SIAM J. Matrix Anal. Appl., 15 (1994), pp. 496–518.

[44] T. COOKLEV AND A. NISHIHARA, *Biorthogonal coiflets*, IEEE Trans. Signal Process., 47 (1999), pp. 2582–2588.

[45] W. DAHMEN, B. HAN, R.-Q. JIA, AND A. KUNOTH, *Biorthogonal multiwavelets on the interval: cubic Hermite splines*, Constr. Approx., 16 (2000), pp. 221–259.

[46] W. DAHMEN, A. KUNOTH, AND K. URBAN, *Biorthogonal spline wavelets on the interval—stability and moment conditions*, Appl. Comput. Harmon. Anal., 6 (1999), pp. 132–196.

[47] W. DAHMEN, A. J. KURDILA, AND P. OSWALD, eds., *Multiscale wavelet methods for partial differential equations*, Vol. 6 of Wavelet Analysis and its Applications, Academic Press, San Diego, CA, 1997.

[48] W. DAHMEN AND C. A. MICCHELLI, *Using the refinement equation for evaluating integrals of wavelets*, SIAM J. Numer. Anal., 30 (1993), pp. 507–537.

[49] I. DAUBECHIES, *Orthonormal bases of compactly supported wavelets*, Comm. Pure Appl. Math., 41 (1988), pp. 909–996.

[50] I. DAUBECHIES, *Ten Lectures on Wavelets*, Vol. 61 of CBMS-NSF Regional Conference Series in Applied Mathematics, SIAM, Philadelphia, 1992.

[51] I. DAUBECHIES, *Orthonormal bases of compactly supported wavelets II: variations on a theme*, SIAM J. Math. Anal., 24 (1993), pp. 499–519.

[52] I. DAUBECHIES AND J. C. LAGARIAS, *Two-scale difference equations II: local regularity, infinite products of matrices and fractals*, SIAM J. Math. Anal., 23 (1992), pp. 1031–1079.

[53] I. DAUBECHIES AND W. SWELDENS, *Factoring wavelet transforms into lifting steps*, J. Fourier Anal. Appl., 4 (1998), pp. 247–269.

[54] G. M. DAVIS, V. STRELA, AND R. TURCAJOVÁ, *Multiwavelet construction via the lifting scheme*, in Wavelet analysis and multiresolution methods (Urbana-Champaign, IL, 1999), Dekker, New York, 2000, pp. 57–79.

[55] D. L. DONOHO, *Nonlinear wavelet methods for recovery of signals, densities, and spectra from indirect and noisy data*, in Different perspectives on wavelets (San Antonio, TX, 1993), Vol. 47 of Proc. Symp. Appl. Math., Am. Math. Soc., Providence, RI, 1993, pp. 173–205.

[56] ———, *De-noising by soft-thresholding*, IEEE Trans. Inform. Theory, 41 (1995), pp. 613–627.

[57] D. L. DONOHO AND I. M. JOHNSTONE, *Ideal spatial adaptation by wavelet shrinkage*, Biometrika, 81 (1994), pp. 425–455.

[58] D. L. DONOHO, I. M. JOHNSTONE, G. KERKYACHARIAN, AND D. PICARD, *Wavelet shrinkage: asymptopia?*, J. R. Stat. Soc. Ser. B, 57 (1995), pp. 301–369. With discussion and a reply by the authors.

[59] G. C. DONOVAN, J. S. GERONIMO, D. P. HARDIN, AND P. R. MASSOPUST, *Construction of orthogonal wavelets using fractal interpolation functions*, SIAM J. Math. Anal., 27 (1996), pp. 1158–1192.

[60] M. DOROSLOVAČKI, *On the least asymmetric wavelets*, IEEE Trans. Signal Process., 46 (1998).

[61] T. R. DOWNIE AND B. W. SILVERMAN, *The discrete multiple wavelet transform and thresholding methods*, IEEE Trans. Signal Process., 46 (1998), pp. 2558–2561.

[62] K. DROUICHE AND D. KALEB, *New filterbanks and more regular wavelets*, IEEE Trans. Signal Process., 47 (1999), pp. 2220–2227.

[63] S. EFROMOVICH, *Multiwavelets and signal denoising*, Sankhyā Ser. A, 63 (2001), pp. 367–393.

[64] S. EHRICH, *Sard-optimal prefilters for the fast wavelet transform*, Numer. Algorithms, 16 (1997), pp. 303–319 (1998).

[65] T. EIROLA, *Sobolev characterization of solutions of dilation equations*, SIAM J. Math. Anal., 23 (1992), pp. 1015–1030.

[66] T. C. FARRELL AND G. PRESCOTT, *A method for finding orthogonal wavelet filters with good energy tiling characteristics*, IEEE Trans. Signal Process., 47 (1999), pp. 220–223.

[67] S. S. GOH, Q. JIANG, AND T. XIA, *Construction of biorthogonal multiwavelets using the lifting scheme*, Appl. Comput. Harmon. Anal., 9 (2000), pp. 336–352.

[68] S. S. GOH AND V. B. YAP, *Matrix extension and biorthogonal multiwavelet construction*, Linear Algebra Appl., 269 (1998), pp. 139–157.

[69] T. N. T. GOODMAN, *Construction of wavelets with multiplicity*, Rend. Mat. Appl. (7), 14 (1994), pp. 665–691.

[70] T. N. T. GOODMAN AND S. L. LEE, *Wavelets of multiplicity r*, Trans. Am. Math. Soc., 342 (1994), pp. 307–324.

[71] R. A. GOPINATH AND C. S. BURRUS, *On the moments of the scaling function ψ_0*, in Proc. ISCAS-92, San Diego, CA, 1992.

[72] B. HAN AND Q. JIANG, *Multiwavelets on the interval*, Appl. Comput. Harmon. Anal., 12 (2002), pp. 100–127.

[73] D. P. HARDIN AND T. A. HOGAN, *Refinable subspaces of a refinable space*, Proc. Am. Math. Soc., 128 (2000), pp. 1941–1950.

[74] ———, *Constructing orthogonal refinable function vectors with prescribed approximation order and smoothness*, in Wavelet analysis and applications (Guangzhou, 1999), Am. Math. Soc., Providence, RI, 2002, pp. 139–148.

[75] D. P. HARDIN AND J. A. MARASOVICH, *Biorthogonal multiwavelets on $[-1,1]$*, Appl. Comput. Harmon. Anal., 7 (1999), pp. 34–53.

[76] D. P. HARDIN AND D. W. ROACH, *Multiwavelet prefilters I: orthogonal prefilters preserving approximation order $p \leq 2$*, IEEE Trans. Circuits Systems II Analog Digital Signal Process., 45 (1998), pp. 1106–1112.

[77] A. HARTEN AND I. YAD-SHALOM, *Fast multiresolution algorithms for matrix-vector multiplication*, SIAM J. Numer. Anal., 31 (1994), pp. 1191–1218.

[78] T.-X. HE, *Biorthogonal wavelets with certain regularities*, Appl. Comput. Harmon. Anal., 11 (2001), pp. 227–242.

[79] C. HEIL AND D. COLELLA, *Matrix refinement equations: existence and uniqueness*, J. Fourier Anal. Appl., 2 (1996), pp. 363–377.

[80] C. HEIL, G. STRANG, AND V. STRELA, *Approximation by translates of refinable functions*, Numer. Math., 73 (1996), pp. 75–94.

[81] P. N. HELLER, *Rank M wavelets with N vanishing moments*, SIAM J. Matrix Anal. Appl., 16 (1995), pp. 502–519.

[82] T. A. HOGAN, *Stability and independence for multivariate refinable distributions*, J. Approx. Theory, 98 (1999), pp. 248–270.

[83] S. JAFFARD, *Wavelet methods for fast resolution of elliptic problems*, SIAM J. Numer. Anal., 29 (1992), pp. 965–986.

[84] L. JAMESON, *On the differentiation matrix for Daubechies-based wavelets on an interval*, Tech. Report ICASE 93-94, Institute for Computer Applications in Science and Engineering, Hampton, VA, 1993.

[85] L. JAMESON, *On the wavelet based differentiation matrix*, J. Sci. Comput., 8 (1993), pp. 267–305.

[86] ——, *On the spline-based wavelet differentiation matrix*, Appl. Numer. Math., 17 (1995), pp. 53–45.

[87] R.-Q. JIA, *Shift-invariant spaces on the real line*, Proc. Am. Math. Soc., 125 (1997), pp. 785–793.

[88] R. Q. JIA AND C. A. MICCHELLI, *Using the refinement equations for the construction of pre-wavelets II: powers of two*, in Curves and surfaces (Chamonix-Mont-Blanc, 1990), Academic Press, Boston, MA, 1991, pp. 209–246.

[89] ——, *On linear independence for integer translates of a finite number of functions*, Proc. Edinburgh Math. Soc. (2), 36 (1993), pp. 69–85.

[90] R.-Q. JIA, S. D. RIEMENSCHNEIDER, AND D.-X. ZHOU, *Vector subdivision schemes and multiple wavelets*, Math. Comp., 67 (1998), pp. 1533–1563.

[91] R. Q. JIA AND Z. SHEN, *Multiresolution and wavelets*, Proc. Edinburgh Math. Soc. (2), 37 (1994), pp. 271–300.

[92] Q. JIANG, *On the regularity of matrix refinable functions*, SIAM J. Math. Anal., 29 (1998), pp. 1157–1176.

[93] Q. JIANG AND Z. SHEN, *On existence and weak stability of matrix refinable functions*, Constr. Approx., 15 (1999), pp. 337–353.

[94] B. R. JOHNSON, *Multiwavelet moments and projection prefilters*, IEEE Trans. Signal Process., 48 (2000), pp. 3100–3108.

[95] J. KAUTSKY AND R. TURCAJOVÁ, *Discrete biorthogonal wavelet transforms as block circulant matrices*, Linear Algebra Appl., 223/224 (1995), pp. 393–413.

[96] F. KEINERT, *Parametrization of multiwavelet atoms*. Appl. Comput. Harmon. Anal., submitted.

[97] ——, *Biorthogonal wavelets for fast matrix computations*, Appl. Comput. Harmon. Anal., 1 (1994), pp. 147–156.

[98] ——, *Raising multiwavelet approximation order through lifting*, SIAM J. Math. Anal., 32 (2001), pp. 1032–1049.

[99] F. KEINERT AND S.-G. KWON, *High accuracy reconstruction from wavelet coefficients*, Appl. Comput. Harmon. Anal., 4 (1997), pp. 293–316.

[100] A. KUNOTH, *Computing refinable integrals — documentation of the program — version 1.1*, Tech. Report ISC-95-02-MATH, Institute for Scientific Computation, Texas A&M Univ., College Station, TX, May 1995. Available at ftp://elc2.igpm.rwth-aachen.de/pub/kunoth/papers/inn.ps.Z.

[101] A. KUNOTH, *Wavelet methods — elliptic boundary value problems and control problems*, Teubner Verlag, 2001.

[102] C. LAGE AND C. SCHWAB, *Wavelet Galerkin algorithms for boundary integral equations*, SIAM J. Sci. Comput., 20 (1999), pp. 2195–2222.

[103] W. LAWTON, S. L. LEE, AND Z. SHEN, *An algorithm for matrix extension and wavelet construction*, Math. Comp., 65 (1996), pp. 723–737.

[104] J. LEBRUN AND M. VETTERLI, *Balanced multiwavelets theory and design*, IEEE Trans. Signal Process., 46 (1998), pp. 1119–1125.

[105] ———, *High-order balanced multiwavelets: theory, factorization, and design*, IEEE Trans. Signal Process., 49 (2001), pp. 1918–1930.

[106] E.-B. LIN AND Z. XIAO, *Multiwavelet solutions for the Dirichlet problem*, in Wavelet analysis and multiresolution methods (Urbana-Champaign, IL, 1999), Vol. 212 of Lecture Notes in Pure and Appl. Math., Dekker, New York, 2000, pp. 241–254.

[107] R. LONG, W. CHEN, AND S. YUAN, *Wavelets generated by vector multiresolution analysis*, Appl. Comput. Harmon. Anal., 4 (1997), pp. 317–350.

[108] W. R. MADYCH, *Finite orthogonal transforms and multiresolution analyses on intervals*, J. Fourier Anal. Appl., 3 (1997), pp. 257–294.

[109] S. G. MALLAT, *Multiresolution approximations and wavelet orthonormal bases of $L^2(\mathbb{R})$*, Trans. Am. Math. Soc., 315 (1989), pp. 69–87.

[110] P. R. MASSOPUST, *A multiwavelet based on piecewise C^1 fractal functions and related applications to differential equations*, Bol. Soc. Mat. Mexicana (3), 4 (1998), pp. 249–283.

[111] P. R. MASSOPUST, D. K. RUCH, AND P. J. VAN FLEET, *On the support properties of scaling vectors*, Appl. Comput. Harmon. Anal., 3 (1996), pp. 229–238.

[112] K. MCCORMICK AND R. O. WELLS, JR., *Wavelet calculus and finite difference operators*, Math. Comp., 63 (1994), pp. 155–173.

[113] Y. MEYER, *Ondelettes sur l'intervalle*, Rev. Mat. Iberoamericana, 7 (1991), pp. 115–133.

[114] C. A. MICCHELLI AND Y. XU, *Using the matrix refinement equation for the construction of wavelets II: smooth wavelets on $[0, 1]$*, in Approximation and computation (West Lafayette, IN, 1993), Birkhäuser, Boston, MA, 1994, pp. 435–457.

[115] ———, *Using the matrix refinement equation for the construction of wavelets on invariant sets*, Appl. Comput. Harmon. Anal., 1 (1994), pp. 391–401.

[116] C. A. MICCHELLI, Y. XU, AND Y. ZHAO, *Wavelet Galerkin methods for second-kind integral equations*, J. Comput. Appl. Math., 86 (1997), pp. 251–270.

[117] J. MILLER AND C.-C. LI, *Adaptive multiwavelet initialization*, IEEE Trans. Signal Process., 46 (1998), pp. 3282–3291.

[118] L. MONZÓN, G. BEYLKIN, AND W. HEREMAN, *Compactly supported wavelets based on almost interpolating and nearly linear phase filters (coiflets)*, Appl. Comput. Harmon. Anal., 7 (1999), pp. 184–210.

[119] H. OJANEN, *Orthonormal compactly supported wavelets with optimal Sobolev regularity: numerical results*, Appl. Comput. Harmon. Anal., 10 (2001), pp. 93–98.

[120] G. PLONKA, *Approximation order provided by refinable function vectors*, Constr. Approx., 13 (1997), pp. 221–244.

[121] G. PLONKA AND V. STRELA, *Construction of multiscaling functions with approximation and symmetry*, SIAM J. Math. Anal., 29 (1998), pp. 481–510.

[122] G. PLONKA AND V. STRELA, *From wavelets to multiwavelets*, in Mathematical methods for curves and surfaces, II (Lillehammer, 1997), Vanderbilt Univ. Press, Nashville, TN, 1998, pp. 375–399.

[123] H. L. RESNIKOFF, J. TIAN, AND R. O. WELLS, JR., *Biorthogonal wavelet space: parametrization and factorization*, SIAM J. Math. Anal., 33 (2001), pp. 194–215.

[124] P. RIEDER, J. GOETZE, J. A. NOSSEK, AND C. S. BURRUS, *Parameterization of orthogormal wavelet transforms and their implementation*, IEEE Trans. Circ. Syst. II, 45 (1998), pp. 217–226.

[125] I. W. SELESNICK, *Multiwavelet bases with extra approximation properties*, IEEE Trans. Signal Process., 46 (1998), pp. 2898–2908.

[126] ———, *Interpolating multiwavelet bases and the sampling theorem*, IEEE Trans. Signal Process., 47 (1999), pp. 1615–1621.

[127] ———, *Balanced multiwavelet bases based on symmetric FIR filters*, IEEE Trans. Signal Process., 48 (2000), pp. 184–191.

[128] W.-C. SHANN AND C.-C. YEN, *On the exact values of orthonormal scaling coefficients of lengths 8 and 10*, Appl. Comput. Harmon. Anal., 6 (1999), pp. 109–112.

[129] L. SHEN AND H. H. TAN, *On a family of orthonormal scalar wavelets and related balanced multiwavelets*, IEEE Trans. Signal Process., 49 (2001), pp. 1447–1453.

[130] L. Shen, H. H. Tan, and J. Y. Tham, *Symmetric-antisymmetric orthonormal multiwavelets and related scalar wavelets*, Appl. Comput. Harmon. Anal., 8 (2000), pp. 258–279.

[131] Z. Shen, *Refinable function vectors*, SIAM J. Math. Anal., 29 (1998), pp. 235–250.

[132] B. G. Sherlock and D. M. Monro, *On the space of orthonormal wavelets*, IEEE Trans. Signal Process., 46 (1998), pp. 1716–1720.

[133] G. Strang, *Linear algebra and its applications*, Academic Press [Harcourt Brace Jovanovich Publishers], New York, 1980.

[134] G. Strang and T. Nguyen, *Wavelets and filter banks*, Wellesley-Cambridge Press, Wellesley, MA, 1996.

[135] V. Strela, *Multiwavelets: theory and applications*, Ph.D. thesis, Massachusetts Institute of Technology, 1996.

[136] V. Strela, *Multiwavelets: regularity, orthogonality, and symmetry via two-scale similarity transform*, Stud. Appl. Math., 98 (1997), pp. 335–354.

[137] V. Strela, P. N. Heller, G. Strang, P. Topiwala, and C. Heil, *The application of multiwavelet filter banks to image processing*, IEEE Trans. Signal Process., 8 (1999), pp. 548–563.

[138] W. Sweldens, *The lifting scheme: a custom-design construction of biorthogonal wavelets*, Appl. Comput. Harmon. Anal., 3 (1996), pp. 186–200.

[139] ———, *The lifting scheme: a construction of second generation wavelets*, SIAM J. Math. Anal., 29 (1998), pp. 511–546.

[140] W. Sweldens and R. Piessens, *Quadrature formulae and asymptotic error expansions for wavelet approximations of smooth functions*, SIAM J. Numer. Anal., 31 (1994), pp. 1240–1264.

[141] J. Tausch, *Multiwavelets for geometrically complicated domains and their application to boundary element methods*, in Integral methods in science and engineering (Banff, AB, 2000), Birkhäuser, Boston, MA, 2002, pp. 251–256.

[142] J. Y. Tham, *Multiwavelets and scalable video compression*, Ph.D. thesis, National University of Singapore, 2002. Available at www.cwaip.nus.edu.sg/thamjy.

[143] J. Y. Tham, L. Shen, S. L. Lee, and H. H. Tan, *A general approach for analysis and application of discrete multiwavelet transform*, IEEE Trans. Signal Process., 48 (2000), pp. 457–464.

References

[144] J. TIAN AND R. O. WELLS, JR., *An algebraic structure of orthogonal wavelet space*, Appl. Comput. Harmon. Anal., 8 (2000), pp. 223–248.

[145] R. TURCAJOVÁ, *Construction of symmetric biorthogonal multiwavelets by lifting*, in Wavelet applications in signal and image processing VII, M. A. Unser, A. Aldroubi, and A. F. Laine, eds., Vol. 3813, SPIE, 1999, pp. 443–454.

[146] P. P. VAIDYANATHAN, *Multirate systems and filter banks*, Prentice Hall, Englewood Cliffs, NJ, 1993.

[147] H. VOLKMER, *Asymptotic regularity of compactly supported wavelets*, SIAM J. Math. Anal., 26 (1995), pp. 1075–1087.

[148] T. VON PETERSDORFF, C. SCHWAB, AND R. SCHNEIDER, *Multiwavelets for second-kind integral equations*, SIAM J. Numer. Anal., 34 (1997), pp. 2212–2227.

[149] M. J. VRHEL AND A. ALDROUBI, *Prefiltering for the initialization of multiwavelet transforms*, IEEE Trans. Signal Process., 46 (1988), p. 3088.

[150] J. R. WILLIAMS AND K. AMARATUNGA, *A discrete wavelet transform without edge effects using wavelet extrapolation*, J. Fourier Anal. Appl., 3 (1997), pp. 435–449.

[151] T. XIA AND Q. JIANG, *Optimal multifilter banks: design, related symmetric extension transform, and application to image compression*, IEEE Trans. Signal Process., 47 (1999), pp. 1878–1889.

[152] X.-G. XIA, *A new prefilter design for discrete multiwavelet transforms*, IEEE Trans. Signal Process., 46 (1998), pp. 1558–1570.

[153] X.-G. XIA, J. S. GERONIMO, D. P. HARDIN, AND B. W. SUTER, *Design of prefilters for discrete multiwavelet transforms*, IEEE Trans. Signal Process., 44 (1996), pp. 25–35.

[154] J.-K. ZHANG, T. N. DAVIDSON, Z.-Q. LUO, AND K. M. WONG, *Design of interpolating biorthogonal multiwavelet systems with compact support*, Appl. Comput. Harmon. Anal., 11 (2001), pp. 420–438.

Index

accuracy p, 29, 143
algorithm
    cascade, 6, 109, 128, 231
    discrete multiwavelet transform, 155
    discrete wavelet transform, 41
antisymmetry, 32, 192
approximation
    coefficient, 29
    multiresolution, 10, 132
    vector, 143
approximation order, 28, 142
    raising by lifting, 206
    symbol factorization, 190
approximation order preserving pre-filter, 160
atom, 202
autocorrelation
    function, 104
    matrix, 226

balanced multiwavelet, 161
band-pass filter, 89
basic regularity, 10, 131
BAT multiwavelet, 241
Bernoulli number, 189
Bezout equation, 74
biorthogonal function, 22, 137
biorthogonality condition, 24, 139
boundary function, 49, 164
boundary handling, 45, 163
    boundary function, 49, 164
    data extension, 46, 163
    matrix completion, 48, 164
building
    multiwavelets, 216
    wavelets, 73

cascade algorithm, 6, 109, 128, 231
    convergence, 109, 231
characteristic function, 248
CL2 multiwavelet, 149, 151, 191, 241
CL3 multiwavelet, 242
co-monic matrix polynomial, 202
coefficient
    approximation, 29
    connection, 95
    recursion, 3, 124
Cohen$(p,\tilde{p})$ wavelet, 85, 203, 240
coiflet, 79, 240
    generalized, 82
completion
    biorthogonal
        cofactor, 214
        QR, 212
    finding dual, 70
    finding wavelet, 69
    orthogonal
        Householder, 211
        projection factors, 209
condition
    basic regularity, 10, 131
    biorthogonality, 24, 139
    E, 108, 131
    E(p), 108, 131
    orthogonality, 17, 136
    Strang–Fix, 30, 145
connection coefficient, 95
constant extension, 47, 164
constant/linear multiwavelet, 124
constrained solution, 223
continuous moment, 26, 140
convolution, 249
correlation, 249

function, 117
cubic Hermite, *see* HC multiwavelet

data extension
    constant, 47, 164
    linear, 47, 164
    periodic, 46, 164
    symmetric, 46, 163
    zero, 46, 164
Daubechies wavelet
    D_2, 6, 72, 77
    D_3, 86
    D_p, 75, 239
    (7,9), 241
    balanced multiwavelet, 163, 242
δ, 248
denoising, 91, 175
DGHM multiwavelet, 126, 142, 144, 158 161, 186, 195, 210, 212, 243
differential equation, 96, 175
dilation factor, 124
discrete moment, 26, 140
discrete multiwavelet transform, 136
    algorithm, 155
    modulation form, 167
discrete wavelet transform, 17
    algorithm, 41
    modulation form, 55
    polyphase form, 57
    two-dimensional, 53
distribution theory, 98, 220
DMWT, *see* discrete multiwavelet transform
downsampling, 41, 156
dual
    function, 22, 137
    lifting, 59, 205
DWT, *see* discrete wavelet transform
dyadic point, 110

equation
    differential, 96, 175
    integral, 96, 175
    matrix refinement, 124
    refinement, 3
errata, 258
existence of ϕ
    as distribution, 98, 220
    in L^1, 100, 223
    in L^2, 102, 225
    pointwise, 110, 231
extended two-scale similarity transform, 185
extension
    constant, 47, 164
    linear, 47, 164
    periodic, 46, 164
    symmetric, 46, 163
    zero, 46, 164

factor
    dilation, 124
    projection, 71
factorization
    lifting, 60, 204
    projection, 71, 199, 202
    spectral, 76
fast
    matrix–vector multiplication, 93
    operator evaluation, 95
fast Fourier transform, 42
filter
    band-pass, 89
    high-pass, 89
    low-pass, 89
formula
    Parseval, 249
    Poisson summation, 249
Fourier transform, 249
Frobenius norm, 252
full rank multiwavelet, 163
function
    autocorrelation, 104
    biorthogonal, 22, 137
    boundary, 49, 164
    characteristic, 248
    correlation, 117
    dual, 22, 137

multiscaling, 132
multiwavelet, 135
orthogonal, 3, 124
refinable, 3
scaling, 10
threshold, 91
wavelet, 16

Hölder continuity, 100
Hölder estimate, 113, 234
Haar wavelet, 4, 16, 71
half-sample symmetry, 47, 164
Hardin–Roach prefilter, 159
hat function, 4
HC multiwavelet, 125, 173, 208, 214, 245
Hermite cubic, see HC multiwavelet
hierarchical preconditioning, 96
high-pass filter, 89
HM multiwavelet, 244
Householder matrix, 252

integral
    equation, 96, 175
integrals
    calculating, 62, 171
    quadrature, 66, 175
Internet wavelet resources, 255
interpolating prefilter, 158

joint spectral radius, 112, 234
JRZB multiwavelet, 244

Kronecker
    delta, 248
    product, 253

L^1, 248
L^1-theory, 100, 223
L^2, 248
ℓ^2, $(\ell^2)^r$, 248
L^2-theory, 102, 225
Laurent polynomial, 250
Leibniz rule, 247
lifting
    multiwavelet, 205

raising approximation order, 206
    wavelet, 59
linear extension, 47, 164
linearly independent shifts, 8, 118, 130, 236
low-pass filter, 89

m-adic point, 231
matrix
    autocorrelation, 226
    Householder, 252
    modulation, 56, 168
    paraunitary, 56, 133
    polyphase, 58, 171
    positive semidefinite, 253
    trace, 253
    transfer, 104, 226
    transition, 104, 226
    TST, 179
matrix completion
    for boundary handling, 48, 164
    polyphase, 209
matrix refinement equation, 124
McMillan degree, 72, 199, 210
modifying
    multiwavelets, 215
    wavelets, 72
modulation matrix, 56, 168
moment
    continuous, 26, 140
    discrete, 26, 140
monomial, 251
mother wavelet, 16
MRA, see multiresolution approximation
multilevel preconditioning, 96
multiplication
    fast matrix–vector, 93
multiplicity of multiwavelet, 124
multiresolution approximation, 10, 132
multiscaling function, 132
multiwavelet, 135
    balanced, 161
    BAT, 241

Index 273

Daubechies, 163, 242
biorthogonal
  HC, 125, 173, 208, 214, 245
  HM, 244
  JRZB, 244
full rank, 163
function, 135
orthogonal
  BAT, 241
  CL2, 149, 151, 191, 241
  CL3, 242
  constant/linear, 124
  DGHM, 126, 142, 144, 158–161, 186, 195, 210, 212, 243
  STT, 243
software, 257
totally interpolating, 159
multiwavelet transform, *see* discrete multiwavelet transform
multiwavelets
  building, 216
  modifying, 215

norm
  Frobenius, 252
normalization, 148
numerical analysis, 93, 175

operator
  fast evaluation, 95
  transfer, 102, 225
  transition, 102, 225
order
  of approximation, 142
order of approximation, 28
orthogonal function, 3, 124
orthogonality condition, 17, 136

paraunitary matrix, 56, 133
Parseval's formula, 249
periodic extension, 46, 164
phase, 57, 169
point values, 35, 147
Poisson summation formula, 249

polynomial
  Laurent, 250
  trigonometric, 250
polyphase
  matrix, 58, 171
  symbol, 57, 170
positive semidefinite matrix, 253
postprocessing, 43, 157
preconditioning
  multilevel, 96
prefilter
  approximation order preserving, 160
  Hardin–Roach, 159
  interpolating, 158
  quadrature, 159
  quasi-interpolating, 160
preprocessing, 43, 157
product
  Kronecker, 253
projection factor, 71
  biorthogonal, 200
  orthogonal, 198
pseudo-identity matrix pair, 202

quadrature prefilter, 159
quasi-interpolating prefilter, 160

recursion coefficient, 3, 124
refinable
  function, 3
  function vector, 124
refinement equation, 3
  matrix, 124
regular two-scale similarity transform, 178
regularity
  basic, 10, 131
resolution 2^{-n}, 12, 134

scale 2^{-n}, 12, 134
scaling function, 10
shifts
  linearly independent, 8, 118, 130, 236

stable, 8, 116, 130
short-term Fourier transform, 20
shrinkage, 91
signal
    compression, 90, 175
    denoising, 91, 175
    processing, 89, 175
singular two-scale similarity transform, 182
smoothness estimate
    L^1, 101, 224
    Hölder, 113, 234
    Sobolev, 106, 228
Sobolev
    smoothness estimate, 106, 228
    space, 105
software, 255
solution
    constrained, 223
    superconstrained, 223
spectral factorization, 76
spectral radius
    joint, 112, 234
stability, 116, 235
stable shifts, 8, 116, 130, 236
Strang–Fix conditions, 30, 145
STT multiwavelet, 243
subband, 89
sum rules, 29, 143
superconstrained solution, 223
superfunction, 145
support, 8, 129
symbol, 5, 127
    polyphase, 57, 170
symmetric extension, 46, 163
symmetry, 32, 191
    half-sample, 47, 164
    whole-sample, 47, 164

thresholding, 91
    hard, 92
    soft, 92
time-frequency transform, 20
totally interpolating multiwavelet, 159

trace of matrix, 253
transfer
    matrix, 104, 226
    operator, 102, 225
transform
    discrete multiwavelet, *see* discrete multiwavelet transform
    discrete wavelet, *see* discrete wavelet transform
    Fourier, 249
    multiwavelet, *see* discrete multiwavelet transform
    short-term Fourier, 20
    time-frequency, 20
    wavelet, *see* discrete wavelet transform
transition
    matrix, 104, 226
    operator, 102, 225
trigonometric polynomial, 250
TST matrix, 179
two-dimensional discrete wavelet transform, 53
two-scale similarity transform
    extended, 185
    regular, 178
    singular, 182

unimodular matrix polynomial, 202
upsampling, 41, 156

vec operation, 252
vector
    approximation, 143
    refinable function, 124
vectorization of matrix, 252

wavelet, 16
    biorthogonal
        Cohen, 85
        Cohen$(p,\tilde{p})$, 203, 240
        Daubechies(7,9), 87, 241
        hat function, 4
    function, 16

orthogonal
   coiflet, 79, 240
   Daubechies D_2, 6, 72, 77
   Daubechies D_3, 86
   Daubechies D_p, 75, 239
   Haar, 4, 16, 71
software, 255
wavelet shrinkage, 91
wavelet transform, *see* discrete wavelet transform
wavelets
   building, 73
   modifying, 72
whole-sample symmetry, 47, 164

z-notation, 250
zero extension, 46, 164